'Bendicsen builds upon his adolescent developmental †¹ ᵉssed in his book, *The Transformational Self*, by offering a ⁓ ⁻o adolescent and early adulthood developmer⁻ descriptive case-study depth he off⁻ further establishes the viability of s\ well-researched text includes ho\ contemporary psychoanalytic the ⌐ₒtanding of the whole person from a bi

Rev. C. Kev ₋ ₐstor, Holy Trinity Parish, Washington, DC

'Harold Bendicsen has written a well-researched and richly synthetic contribution to our understanding of development, case formulation and approaches to treatment, incorporating the multiple frames of reference from philosophy, to subjectivity to neuroscience available in our current state of the art. This creative scholarly work is a must for academicians and clinicians alike.'

Charles M. Jaffe, MD, Training and Supervising Psychoanalyst, Chicago Psychoanalytic Institute; Assistant Professor of Psychiatry, Rush University Medical Center, Chicago, Illinois

'The concept of a "developmental algorithm" is a dazzling use of current modern metaphor for the analysis of human processes. Bendicsen's metaphor creates a sufficient connection and platform for the discussion and integration of non-linear conceptualizations and theories of human development. A developmental algorithm has the possibility of becoming the accessible key to unlock the heavily encrypted psychoanalytic theories of development that have been siloed away for decades. This metaphor allows for multiple forces, both internal and external, to be observed in and to the "self" on its journey from birth to death. Bendicsen's multi-modal description of the necessary processes for the *Regulation of the Self* provides a vantage point to the multiple processes at work.'

William Gieseke, PhD, Fundamentals Year Faculty, Chicago Psychoanalytic Institute; Faculty, Child and Adolescent Psychoanalytic Psychotherapy Program, Chicago Psychoanalytic Institute

'In the *Regulation of the Self*, Harold Bendicsen provides us with a comprehensive, scholarly and contemporary integration of psychoanalysis and neuroscience. Drawing on his own clinical work and extensive study of the literature he shows us where psychoanalysis is today, how we got here, and how to use this knowledge to help our patients successfully become themselves.'

Neal Spira, MD, Training and Supervising Analyst, Dean, Chicago Psychoanalytic Institute

PSYCHOANALYSIS, NEUROSCIENCE AND ADOLESCENT DEVELOPMENT

Psychoanalysis, Neuroscience and Adolescent Development: Non-Linear Perspectives on the Regulation of the Self explores how psychoanalysis can combine its theoretical perspectives with more recent discoveries about neurological and non-linear developmental processes that unfold during the period of puberty to young adulthood, to help inform understanding of contemporary adolescent behaviours and mental health issues.

With the powerful impact of neuroscience research findings, opportunities emerge to create a new paradigm to attempt to organize specific psychoanalytic theories. Neurobiological regulation offers such an opportunity. By combining elements of domains of compatible knowledge into a flexible explanatory synergy, the potential for an intellectually satisfying theoretical framework can be created. In this work, Harold Bendicsen formulates a multidisciplinary theoretical approach involving current research and drawing on neuroscience to consider the behaviour regulation processes of the mind/brain and the capacities and potential it brings to understanding the development of adolescents and young adults.

Psychoanalysis, Neuroscience and Adolescent Development advances Bendicsen's study of adolescence and the transition to young adulthood, begun in *The Transformational Self*. It will be of great interest to psychoanalysts and psychoanalytic psychotherapists, as well as psychologists, clinical social workers, psychiatrists and counsellors.

Harold K. Bendicsen, LCSW, BCD, is a Clinical Social Worker who maintains a private psychotherapy practice in Elmhurst, Illinois. He holds a certificate in Child and Adolescent Psychoanalytic Psychotherapy from the Chicago Psychoanalytic Institute. He is Adjunct Professor at Loyola University Chicago School of Social Work and a member of the faculty of the Child and Adolescent Psychotherapy Training Program at the Chicago Psychoanalytic Institute. With Joseph Palombo and Barry Koch he has co-authored *Guide to Psychoanalytic Developmental Theories* and authored *The Transformational Self: Attachment and the End of the Adolescent Phase*.

PSYCHOANALYSIS, NEUROSCIENCE AND ADOLESCENT DEVELOPMENT

Non-Linear Perspectives on the Regulation of the Self

Harold K. Bendicsen

Routledge
Taylor & Francis Group
LONDON AND NEW YORK

First published 2019
by Routledge
2 Park Square, Milton Park, Abingdon, Oxon OX14 4RN

and by Routledge
52 Vanderbilt Avenue, New York, NY 10017

Routledge is an imprint of the Taylor & Francis Group, an informa business

© 2019 Harold K. Bendicsen

The right of Harold K. Bendicsen to be identified as author of this work has been asserted by him in accordance with sections 77 and 78 of the Copyright, Designs and Patents Act 1988.

All rights reserved. No part of this book may be reprinted or reproduced or utilised in any form or by any electronic, mechanical, or other means, now known or hereafter invented, including photocopying and recording, or in any information storage or retrieval system, without permission in writing from the publishers.

Trademark notice: Product or corporate names may be trademarks or registered trademarks, and are used only for identification and explanation without intent to infringe.

British Library Cataloguing-in-Publication Data
A catalogue record for this book is available from the British Library

Library of Congress Cataloging-in-Publication Data
A catalog record has been requested for this book

ISBN: 978-0-367-13494-5 (hbk)
ISBN: 978-0-367-13496-9 (pbk)
ISBN: 978-0-429-02681-2 (ebk)

Typeset in Bembo
by Newgen Publishing UK

To Joseph A. Walsh

CONTENTS

Preface	*xiii*
Acknowledgements	*xv*
Introduction	1
1 The succession of psychoanalytic theories from drive theory to neuropsychoanalysis	**3**
The ever shifting state of our theorizing	3
Drive theory and beyond	6
Neuropsychoanalysis	12
Anticipating a neuropsychoanalytic metapsychology	15
Notes	18
2 Regulation theory: toward a new paradigm	**22**
Two perspectives on brain organization – the triune brain and the social brain	24
Neurobiological regulation	27
A suggested epigenetic framework for regulatory processes	28
Types of regulatory systems in the social brain	29
Self-state regulation	31
The biopsychosocial model	32
Tronick's Mutual Regulation Model	33
Regulation and resilience	37
Schore's neuroendocrinology regulatory hypothesis of boys at risk	39
Note	40

3	Case example	42
	Myles' journey – phase one	43
	Myles' journey – phase two	44
	Month 1 to month 17	*44*
	Discussion	47
	Issues in the two phases	*47*
	Month 17 to month 49	*49*
	Discussion	*53*
	Month 49 to month 56	*54*
	Discussion	*55*
	Month 56 to month 62	*56*
	Discussion	*57*
	Note	58
4	Alternative developmental model thinking	59
	Lapham's wound	59
	The Transformational Self	62
	Toward a compatible set of interlocking theories	64
	A definition of regulation theory	65
	A definition of a developmental algorithm and the rationale for the selection of criteria in a developmental algorithm	66
	The linkages among self psychology, intersubjectivity theory and relational psychoanalysis	68
	Theoretical considerations	80
	Notes	85
5	Case formulation from a regulation theory perspective	88
	Case formulation using a developmental algorithm	88
	Modern metaphor theory	*88*
	Attachment theory	*90*
	Self psychology with intersubjectivity theory and relational psychoanalysis	*93*
	Supportive relationships	96
	Cognitive theory	*98*
	Contemporary psychoanalytic developmental psychology	*101*
	The developmental model	101
	Developmental processes	111
	Complexity theory or non-linear dynamic systems theory	*118*
	Neurobiology with narrative theory	*121*
	Notes	124

6 Desiderata: further thoughts on case formulations	128
From a traditional psychodynamic/psychoanalytic self-psychological perspective	129
Part one – summarizing statement	129
Part two – description of nondynamic factors	130
Part three – psychodynamic explanation	130
Part four – predicting responses to the therapeutic situation	131
From complementary orientations	132
Psychodynamic explanation – the social brain perspective	132
From a complementary non-linear dynamic systems perspective	134
Psychodynamic explanation – non-linear dynamic systems perspective	136
Note	137
7 Treatment efficacy	138
The treatment of schizophrenia	138
Note	141
8 Some thoughts on the adolescent passage, emerging adults and millennials	142
9 Synopsis	148
Notes	157
Appendix I: Critical thinking mental health decision-making flow chart	*158*
Appendix II: Theories of alcoholism	*160*
References	*164*
Index	*183*

PREFACE

Professional exposure to my ideas on this subject can be said to have begun with a presentation in a clinical newsletter. An abridged version of this monograph was published in the *Illinois Society for Clinical Social Work Newsletter*, Spring, 2014, under the title of "The Argument for a Compatible Set of Interlocking Theories as Applied to a Case Example" (ISCSW, P.O. Box 2929, Chicago, IL 60690–2929). I want to thank Ruth Sterlin, President of the ISCSW, and her successor, Eric Ornstein, for their encouragement and persistence in helping me compress these expansive concepts into bite size morsels.

In addition, this unabridged monograph won the 2014 Yellowbrick Foundation Emerging Adult Paper Prize on June 20, 2014 sponsored jointly by The Yellowbrick Foundation (1560 Sherman Avenue, Suite 400, Evanston, IL 60201) and the Chicago Institute for Psychoanalysis (122 South Michigan Avenue, Suite 1300, Chicago, IL 60603). A special thank you goes out to both Jesse Viner, MD, Founder and Executive Medical Director of Yellowbrick and Paul Holinger, MD, Dean, of the Chicago Institute for Psychoanalysis for their warm receptivity to my ideas and their vigorous encouragement to present them to a wider audience. At Yellowbrick, I also want to thank both David Daskovsky, Senior Psychologist, and Michael Loseff, Staff Psychologist, for arranging and facilitating the fruitful exchange of ideas amongst staff and this author.

The underlying conviction of this writing project can be framed as an effort to integrate the promise that neuroscience research findings offer in informing new theoretical hypotheses in understanding human motivation. The presentation and arrangement of material is geared to lead the reader from global theoretical considerations in the psychoanalytic enterprise to its intersection with neuroscience. In this monograph I propose a developmental algorithm, in the context of operationalizing regulation theory, to acquaint the reader with a framework to use as neuropsychoanalytic propositions continue to emerge. The ability of psychoanalytic theorists to synthesize their viewpoints by producing a unifying theory, a "*Weltanschauung*" (Freud, 1933), if you will, is clearly out of reach, and is not the point of this writing project.

In Chapter One, I trace the vicissitudes of psychoanalytic theory formation and conclude with the emergence of neuropsychoanalysis.

In Chapter Two, I review brain organization and the need for understanding the overarching biopsychosocial regulation principles of human functioning.

In Chapter Three, I proceed by presenting a case example of a late adolescent/young adult male experiencing severe dysregulation and the struggle to stabilize his life. I include discussions to the case record to give the reader a glimpse of my thinking on the biopsychosocial dynamics, the challenges ahead and the successful prospects for this resilient individual.

In Chapter Four, I discuss two pathways to adulthood, one fairly straightforward and the other ambiguous. I suggest that those emerging adults who develop the capacities for self-reflection, self-regulation and the delay of gratification have an opportunity for an integrated self. This new self-state, a metamorphosis, if you will, may be brought about by the *appearance of self-referencing metaphor that can evolve into self-regulating metaphor* which can channel the identifications and potentials of the late adolescent into a stable young adulthood. I also reflect on theoretical models.

In Chapter Five, I introduce the concept of a developmental algorithm as an alternative way to explain and, otherwise, clarify, human processes and case dynamics. This construct is grounded in theoretical pluralism.

In Chapter Six, I display a variety of complementary case formulations including traditional self psychology, and informed by neuroscience, a social brain perspective and a non-linear dynamic systems viewpoint, all highlighting the compatibility of employing psychoanalytic and nonpsychoanalytic formulations.

In Chapter Seven, I review contemporary consensus treatments for spectrum schizophrenia.

In Chapter Eight, I reflect on thoughts about the passage through adolescence from two major coming-of-age perspectives, the unambiguous rite of passage and the amorphous developmental drift into young adulthood. I also consider the validity of the emerging adulthood developmental concept and I discuss the Millennials as the target group and their impact on extending the adolescent phase.

In Chapter Nine, I conclude with a summary highlighting the journey from a splintered psychoanalytical theoretical field to one informed by neuroscience and nontraditional perspectives with the ensuing prospects for more coherent elaboration of therapeutic frameworks. I maintain that the intersection of neuroscience, non-linear dynamic systems theory and contemporary psychoanalytic thinking will offer powerful explanatory synergies to clarify motivational forces and illuminate the often opaque human condition.

ACKNOWLEDGEMENTS

I am most appreciative to the Yellowbrick Foundation for their interest in my ideas and their enthusiastic assistance in exposing these ideas to a wider audience. Also, I want to express my grateful appreciation to the following colleagues for their collaborative thoughts and comments in the preparation of this monograph: Joe Palombo, Alan Levy, Barrie Richmond, Rita Sussman, Joseph A. Walsh, Ed Kaufman, Barry Childress, William Gieseke, Barbara Alexander Neal Spira, Charles Jaffe, Virginia Barry and Matthew M. Malerich. My monthly study group, consisting of Debbie Barrett, Middy E. Fierro, Kathleen M. O'Connor, Char Slezak, Julie Caron-Sims and especially Rosalie Price, J. Colby Martin and Bernadette Berardi-Coletta, have again provided valuable feedback. They have challenged my conceptualizations and made sure my intellectual leaps were appropriately tethered. A special thank you to Renee Siegel for grounding the case of Myles in a more coherent self-psychological framework. I am deeply indebted to Joe Palombo for clarifying the philosophical underpinnings of two theoretical traditions, intersubjectivity theory and relational psychoanalysis, and to Alan Levy, and especially Carol Ganzer, for explicating the nature of the interplay between these two traditions. Considerable recognition goes to Andrew Suth for his careful review of my theoretical formulation and his generous contributions in pointing out relevant cognitive research literature. Credit also to Rosalie Price and J. Colby Martin for questioning my ideas and pushing me to say what I mean with clarity. I also want to thank my former student, Kevin McMahon, who invited me to his normal human development class to expose his students to the explanatory power of the regulation hypothesis. In terms of clarifying the nature of the term "developmental algorithm," I am indebted to the following students in a psychotherapy class (February 9, 2018) at the Chicago Institute for Psychoanalysis in the Child and Adolescent Psychoanalytic Psychotherapy Program: Neha Lodha, Jessica Johnson and Simona Bommarito. I am grateful to Rev. Kevin Gillespie for his assistance in clarifying the theological nature of a self-referencing metaphor attributed to Jesus of Nazareth. Kudos to Ellen Fechner for her formatting assistance. And once again, sincere thanks to my always reliable and conscientious proof readers: Elizabeth Bendicsen, Dorothy Valintis, Kathleen Bendicsen and Gary Michelsen.

INTRODUCTION

Psychoanalytic theory construction can result in magnificent architecture offering compelling perspectives on human experience. Closer examination of the positioning of these structures, however, reveals a fluid landscape containing sinkholes and quicksand threatening to befuddle the traveler who searches for the best vantage. Competing tribes commit vast resources to advertise their patch as the best place to sojourn. With the powerful impact of neuroscience research findings, opportunities emerge to create a new paradigm to attempt to coordinate these tribes. Neurobiological regulation offers such an opportunity. By combining elements of domains of compatible knowledge into a flexible explanatory synergy, the potential for an intellectually satisfying theoretical framework can be created. In this project I will expand on two themes first articulated in *The Transformational Self: Attachment and the End of the Adolescent Phase* (2013): 1) I will clarify the nature of regulation processes placing selected ones in a developmental context and 2) I will demonstrate the theoretical usefulness of using a set of interlocking theories to form a developmental algorithm, a formula to operationalize regulation theory and illustrate its therapeutic efficacy through an extended case example.

This monograph is the second part of an extended study on the nature of the contemporary self. *Psychoanalysis, Neuroscience and Adolescent Development: Non-Linear Perspectives on the Regulation of the Self* is a continuation of the conceptual work started in the *Transformational Self: Attachment and the End of the Adolescent Phase*. In *The Transformational Self* I presented a fresh perspective on the age-old question of "When does adolescence end?" In *Psychoanalysis, Neuroscience and Adolescent Development: Non-Linear Perspectives on the Regulation of the Self* I examine the properties of the dynamic self-structure contextualized in an ever-changing developmental life cycle framework. Emphasis is placed on one of the most complex phases, the transition from late adolescence into young adulthood. In this project, the reader is invited to shift from a linear, epigenetic process to a non-linear, probabilistic epigenetic perspective in understanding change and growth. The self never exists as an isolated entity, but rather is embedded in an attachment web of mutually influencing relationships. The biopsychosocial-spiritual forces continuously impinging on the self-structure function in a matrix of subjective meaning-making relationships with the ceaseless ebb and flow of personal narrative construction and reconstruction.

The regulation processes are manifold and include all biopsychosocial-spiritual forces in interaction in a human systems context. As development unfolds, these processes need to be adjusted and modulated in different modes of coordination as the organism adapts to change and adversity. Key to the survival of the organism is the need to remain in balance (self-righting or adaptive capacity) and resilient as its constitution undergoes continual reconfiguration. The focus of this monologue will be on explicating the dynamics associated specifically with the developmental passage into emerging adulthood. Traditionally, the passage was assessed as the measure by which a collection of tasks were addressed and completed. Clinical professionals trained in today's *zeitgeist* have moved beyond these parameters. I will shift the discussion from tasks to be accomplished to the set of processes that need to be understood. The fundamental process is the emergence of the Transformational Self, a phase-specific dimension of the neural self, a reconfigured, coherent state of mind. The Transformational Self is a way to link the 1) maturation of the prefrontal cortex and its improved interconnectivity with 2) enhanced executive functioning and 3) the possibility for self-referencing metaphor to potentiate new identifications. In other words, this project explores the means by which self-referencing metaphor evolves into self-regulating metaphor and becomes the pathway to personal metamorphoses.

> Then the noble hero, looking forth upon the wide water spread before his eyes, pointed with his finger and said: "What place is that? Tell me the name which that island bears. And yet it seems not to be one island." The river god replied: "No, what you see is not one island. There are five islands lying there together; but the distance hides their divisions."
>
> (Ovid, *The Metamorphoses, Book VIII*)

1
THE SUCCESSION OF PSYCHOANALYTIC THEORIES FROM DRIVE THEORY TO NEUROPSYCHOANALYSIS[1]

The ever shifting state of our theorizing

The 125-year history of the psychoanalytic movement is littered with the record of schisms, a testament to its inability to accommodate emerging theoretical positions. It is so much a part of psychoanalytic history that it has become a distinguishing feature of its core nature. In this fertile area of potential cross-fertilization of ideas, the journey is not infrequently through or around unfriendly, even hostile, camps of differing opinion. What accounts for this theoretical balkanization? Is it because of the powerful personal investments in our ideational creations that understands contrary opinions as a diminishment of our narcissistic creativity (Kohut, 1966), which is too painful to tolerate? Surely, in the exchange of ideas, this state of affairs is seen in virtually all walks of life. But why should this be so endemic in psychoanalysis?

Goldberg (2013), in discussing the diversity and pluralism of psychoanalytic theorizing, has framed the issue in unambiguous terminology.

> One paper stands out in my mind as especially incomprehensible because of a variety of symbols and words that seems almost to resemble a secret code. A colleague explained to me the particular theory that these symbols represented, and essentially the theory made good sense, once an effective translation could be accomplished. No doubt the author of this particular paper felt that he or she was writing to like-minded members of his or her group and so gave little mind to those of us who might fall outside of such membership. When Freud wrote about the "narcissism of minor differences," he indicated that groups were bound close together and so were able to keep other groups at bay by these allegiances, which were fueled by aggressive impulses (Freud, 1910, p. 199). A good deal of anger and irritation is often felt in reading of clinical cases or non-clinical papers which pursue a vocabulary

4 The succession of psychoanalytic theories

of terms which are idiosyncratic and may seem to exclude a subset of aliens. Thus Freud's explanation seems quite correct.

(Goldberg, 2013, p. 1)

The tendency to form exclusive secret theoretical clubs runs the risk of further polarizing and fragmenting the greater psychoanalytic community, not to mention the academic community. "Our diverse schools of psychoanalysis are more devoted to defending borders than to an open exchange of information" (p. 12). Optimistically, envisioning the possibility of an evolving collaborative process, Goldberg concludes, "Once we appreciate the value of other analytic efforts, we may be able to establish and utilize a common vocabulary as well as a common vision of effectiveness" (p. 12).

Goldberg's sentiments are echoed by Barratt (2013) who uses the term "theoretical bedlam" to characterize the intense psychoanalytic theoretical fragmentation.

> Indeed, the history of the psychoanalytic movement through the 20th century provides an appalling spectacle of in-groups and tightly knit clubs fighting with each other, of internecine rivalries, backbiting, and name-calling that is only thinly veiled in professional courtesy, of rampant authoritarianism, of politically motivated expulsions and the formation of counter-organizations, and thus of proliferative organizational splitting.
>
> (In O'Loughlin, 2016, p. 5)

Over twenty years ago, P. Tyson et al. had already taken the argument further in reflecting on expansionist tendencies:

> The point here is simply that each contending theoretical perspective in psychoanalysis, however narrowly defined its original base of inquiry and intervention within the domain of psychoanalytic application, is soon broadened to become a proclaimed superior metapsychology and intervention system for all those whose mental and emotional distress bring them within the psychoanalytic professional orbit.
>
> (Tyson, Tyson and Wallerstein, 1990, p. x)

Cooper (2008), reflecting on the entrenched "plurality of authoritarian orthodoxies" (p. 250), wrote, "Among many psychoanalysts an innovative idea is more likely to be regarded as an assault rather than an interesting opportunity" (p. 241). In addition,

> Our aesthetics is important in determining how we think analytically. Some of us are more interested in storytelling – a hermeneutic view; some of us are more interested in archaeology – reconstructing the past; some of us are more interested in architecture – providing a structure to human life. Our aesthetic

preferences — our 'taste' and sensibilities — are very difficult to change, even in the face of "facts".

(p. 250)

Hinshelwood (2013) echoes these positions. "Psychoanalytic schools narcissistically guard the group's knowledge and set it against others with a hostility approaching siege mentality. This leads to further retreat into our separate groups, each marked by its conceptual convictions" (p. 8). So the more our theories proliferate, the more isolationist our theoretical camps, with their different thought styles, seem to become. It is hoped that regulation theory, with its emphasis on a set of multidisciplinary, interlocking theories, can offer an opportunity to build flexible bridges among these fixed conceptual positions.

As adherents of particular theories lay claim to possessing the truth, how do we know what truth is in psychoanalysis? Hanly (1990, in Cooper 2008) suggests that two ways of assessing truth are possible and described correspondence and coherence theories of truth.

> Correspondence assumes that an assertion is true if it accurately describes something that is actually "out there." Coherence assumes that a statement is as close as we can get to truth if it fits within a general scheme of how one sees the world.
>
> (p. 245)

I will advance my regulatory hypothesis using a coherence position argument in this monologue.

As a heuristic device, allow me to compare and contrast the "hard" v the "soft" sciences. If psychoanalytic theorizing is fragmented, it seems to be less the case in scientific fields, the so-called "hard sciences," such as physics, chemistry and biology. In those endeavors theory building (i.e., theory that is settled and accepted) rests on the accumulation and deepening of data in a progressive, collaborative way. Investigations proceed by building on the concepts and findings of previous theory and research. A sharper boundary exists between the explanatory systems and empirical phenomena. A physical theory, for example, need only account for physical phenomena without obligation to account for competing theories. Sensitivity to the history of previous discoveries is keen (Palombo, 1991). However, such a sharp division is not always the case as we see in the discovery of quantum mechanics where theoretical disagreements and arguments consumed enormous personal energy on the part of the adherents (Cassidy, 1992, pp. 226–246).

In so-called "soft sciences," such as anthropology, sociology, personality psychology, research is carried on in limited domains without much attention paid to competing theories. Consequently, personality psychology is largely a noncumulative affair (Palombo, 1991 in Bendicsen, 2013). In the absence of a grand unifying theory, a *Weltanschauung* (Freud, 1933, p. 158), to borrow a term characterizing Freud's intellectual quest, competing personality theories abound with their hypotheses assuming the certitude of religious dogma. Assuming Palombo's dichotomy is essentially correct, does the psychoanalytic enterprise not thrive on collaborative theory building because its foundation in biology is too tentative and, consequently, it inherently fits better into the category of the "soft sciences?" I am reminded (in personal communications with J. Colby Martin on September 8, 2013 and William Gieseke on October 7, 2013) that it is a fallacy of reductionism to believe that the study of the nature of human experience, the domain of psychoanalysis, can be made more "real" by attaching it to the biological sciences. As has been often observed, many detractors of psychoanalysis consider psychoanalysis to have more in common with religion than science. Jung, perhaps Freud's most significant detractor, on his 1909 visit to the United States in an apparently off-hand comment, took such a position. Disputing Freud's claim that psychoanalysis should be considered among the sciences, Jung said, "Psychoanalysis is not a science but a religion" (Gay, 1988, p. 239).

Drive theory and beyond

Drive theory developed out of Freud's frustrated attempt to fashion a new psychology based on the evolving understanding of the organization of the nervous system, the exciting discovery of the neuron and the designation of this brain cell by the name of "neuron," all by 1891 (Freud, 1895). By 1897 the Neuron Doctrine with four tenets had been formulated: 1) the neuron is the fundamental structure and functional unit of the nervous system; 2) neurons are discreet cells which are not continuous with other cells; 3) the neuron is composed of three parts – the dendrites, axon and cell body; 4) information flows along the neuron in one direction (from the dendrites to the axon via the cell body) (Costandi, 2006, p. 18).

A confluence of formative experiences in the 1880s, when Freud was in his 30s, fostered a creative synergy which led Freud to move away from laboratory research in neurophysiology to the office practice of neurology. Freud's study of hypnosis with Jean Martin Charcot in Paris in 1885–1886 at the Salpetriere led Freud to abandon the German method of Continental neurology in which the anatomical and physiological parts of the clinico-anatomical equation were stressed. "Clinical material served the secondary purpose of demonstrating and confirming existing anatomical and physiological theory" (Kaplan-Solms and Solms, 2002, p. 11). After Broca's and Wernicke's seminal discoveries in the aphasias during the 1860s and 1870s the race was on to discover and identify localization of other mental functions, normal and abnormal. Broca pinpointed the function of expressive language in the left hemisphere at the junction of the parietal and temporal lobes and Wernicke located the function of receptive language in the left hemisphere in the temporal lobe.

In the French school of Continental neurology, the emphasis fell upon the clinical side of the equation. The aim was to develop correlations between external manifestations and signs and the interior of the disease process. This was the origin of the concept of clinical syndromes. Freud became expert on the clinico-anatomical method but grew to feel its limitations and that the future of neurology lay in expanding the potential of the clinical part of the equation.

With Freud's return from Paris his interest began to shift from the localization of functions in the anatomical study of the aphasias to the non-localization of functions in the clinical psychological study of the neuroses (using hypnosis), especially hysteria and neurasthenia. Over the next decade Freud, still a neurologist, expanded his work on the aphasias (Freud, 1891) to include the neuroses (Freud and Breuer, 1895). While the German and French schools of neurology generally complemented each other well in the study of the aphasias, a divide appeared in the study of the neuroses. The French school was less interested in explaining the clinical pictures on the basis of existing theory. Rather its emphasis lay in identifying, classifying and describing them; so French neurology was first and foremost a nosology in the broadest sense of classifying diseases (Kaplan-Solms and Solms, 2002, p. 12). Freud made two groundbreaking observations linking the aphasias and the neuroses with the non-localization of functions. First, "psychological faculties break down according to the logic of *their own functional laws,* not according to the laws of cerebral anatomy." And second, "psychological functions are never destroyed, by localized brain lesions … Rather, they are distorted and changed in dynamic ways that reflect a mutual interdependence with other faculties" (p. 17). Here we see the seeds of a formative metapsychology, the underpinning to later psychoanalytic thinking.

In addition to the discovery of neurons and Freud's shift to French neurology, a third factor that contributed to the formation of drive theory were the close collaborations with Wilhelm Fliess, a Berlin physician, during the years from 1887 to about 1900 and a parallel collaboration with Josef Breuer, a fatherly mentor, physician and co-author of *Studies on Hysteria* (1895d), which lasted from 1887 to about 1898. (See Palombo, Bendicsen and Koch, pp. 6–7 for details and perspectives on these significant partnerships.) Freud drew on each working friendship to produce conceptual yields which formed the underpinnings to drive theory. From the work with Fliess, Freud clarified his thinking on the ideas of "reality testing, the formal distinction between primary and secondary processes, and the wish fulfillment of dreams" (Preface to the "Project for a Scientific Psychology" in the *Standard Edition,* Vol. I, p. 284 and Palombo, Bendicsen and Koch, p. 5). From the work with Breuer, Freud developed the ideas of free association, resistance, repression, symptom formation, conversion, transference, countertransference, symbolic content and unconscious conflict (Berzoff, Flanagan and Hertz, 2008, p. 27). These concepts formed Freud's well-known metapsychological framework: the economic hypothesis (1895), the topographical hypothesis (1895), the dynamic hypothesis (1895) and the genetic (or psychosexual) hypothesis (1905). These four hypotheses became the foundation of drive theory. The term drive, for example, the libidinal drive, refers to the psychological manifestation of a biological instinct. Drive theory became the wellspring for all the variations of psychoanalytic theory to follow. From this point forward all subsequent psychologies that lay claim to being psychoanalytic have at their base, to one degree or another, an abiding belief in the dynamic unconscious, developmental model thinking and transference and countertransference. (See either the Palombo, Bendicsen and Koch, pp. 1–45, or the Berzoff, Flanagan and Hertz, 2008, pp. 17–47 sources for a further elaboration of drive theory.)

While Freud's collaboration with Breuer (*Studies on Hysteria*, 1895) produced an immediate and intense interest in the subject of psychoanalysis, the more removed collaboration with Fliess (*Project for a Scientific Psychology*, 1895) had to wait until 1954 and the English translation for the discipline of dynamic neuropsychology to become reinvigorated. It is known that Freud wrote a large section of his *Project for a Scientific Psychology* (1895) and sent it unfinished to Fliess for review. Essentially, Freud could not align his main problem, that of psychological repression, with the primitive neuron theory of his day (Sulloway, 1979, pp. 123–130). Freud then asked Fliess to destroy it, which Fliess fortunately did not.

To sum up this section,

> Freud carried over from neurology into psychology a basic method – namely the clinical-descriptive method of Charcot (or the method of syndrome analysis, as it later came to be known) – and a basic conceptualization of (functional, author's insertion) mind-brain relationships – namely the anti-localization or dynamic conceptualization – which gives pride of place to psychological methods of analyzing psychological syndromes *regardless of whether or not they have an organic etiology.*
>
> (Kaplan-Solms and Solms, 2002, p. 25)

In much briefer fashion I now turn to an enumeration of the record of post drive theory varieties of psychoanalytic theorizing.

Psychoanalytic theoretical insularity may be said to have begun with Freud's bitter experience with the theoretical disagreements related originally to his study group and his inability to shape doctrinal homogeneity. This led to the subsequent separation of Stekel, Adler, Jung and later Rank from the orthodox psychoanalytic community and their enforced ostracism through the Secret Committee formed in 1912 (Grosskurth, 1991). This pathway formed an inflexible template for dealing with differences. Theoretical innovations could be accepted, of course, but they had to be with Freud's blessing.[2] In this regard see the contributions from Abraham (1927), Ferenczi (1988) and Hartmann (1939). The Committee was committed to safeguarding Freud's core theory involving the dynamic unconscious, repression and infantile sexuality as embodied in his two Herculean, interlocking, intellectual syntheses: the model of the mind synthesis (1900) and the psychosexual synthesis (1905) (Makari, 2008 in Palombo, Bendicsen and Koch, 2009, p. 13). But the "centre" would not hold and by 1927 the work of the Committee was dissolved, its affairs passed over to the International Association (Grosskurth, 1991, pp. 17, 117, 192; Hinshelwood, 2013, p. 8).

The process of schism continued through the debates between Melanie Klein vs. Anna Freud. The schism led to the famous three-way split among Kleinian followers, Anna Freud adherents and the so-called "independent" or middle school of non-Kleinian theorists, which became the object relations school (Mitchell and Black, 1995; Summers, 2006). It moved to the controversy as to whether or not the instincts were fundamentally discharge seeking or object seeking (Fairbairn, 1941; 1952). Freud's four-part structural/tripartite theory (id, ego, superego/ ego ideal) dominated thinking until the ego's defensive and adaptive roles expanded; structural theory was absorbed into ego psychology. Then the ego psychologists, held dominion in psychoanalytic theorizing from the late 1930s to the mid-1970s, found their position eroded and subsequently supplanted by self psychology (Mitchell and Black, 1995; Pine 1990). Kohut's monumental theoretical achievement and its subsequent schismatic effects are captured by Schechter (2014) in her intimate historical overview of the legendary Chicago Institute for Psychoanalysis (CIP). (In 2018 the CIP was renamed the Chicago Psychoanalytic Institute.) Detailed is the impact of the first schism, that of the school of Lionel Blitzsten who advocated for an orthodox, narrow, intrapsychic closed system view of psychoanalytic thinking and practice versus those of the founder, Franz Alexander, who advanced modifications in technique and outreach to the larger social milieu. In the struggle between these two apostolic giants, the question of authority was paramount. The aftermath set the stage for the decades-long bitterness and viciousness over who would be labeled a theoretical gatekeeper, visionary or apostate. The current debate within the Chicago Psychoanalytic Institute over its larger community relevance, and even its survival as a school of psychology and mental health discipline, traces its roots directly to this complete inability to accommodate differences. Unfortunately, the theoretical balkanization continues.

Led by Spitz (1945a; 1945b; Spitz and Wolf, 1946; 1965), the infant observation researchers entered the theoretical playground. Mahler's infant work was considered groundbreaking, but after only ten years, Mahler was replaced by Stern (Mahler, Pine and Bergman, 1975; Stern, 1985, Applegate, 1989). Efforts to synthesize divergent intrapsychic theories, such as those of Kernberg (1976b; Palombo, Bendicsen and Koch, 2009) and Tyson, Tyson and Wallerstein (1990), have been few and not widely accepted. With the consolidation of the self as the unit of analysis (i.e., the search for understanding the nature of a resilient self-structure) and the ascendency of the social constructivists, the deductive theorists (those who impose presuppositions on data) are being marginalized by the hermeneutic interpretationists (those who use theory as a heuristic, formulating suppositions, and take as the unit of analysis the experience of the subject) (Summers, 2013).

In 1980 the publication of the *Diagnostic and Statistical Manual of Mental Disorders* (DSM-III) heralded a sea change in the manner by which disorders are classified. In so doing it further marginalized psychoanalytic theorizing and generated extensive controversy. By then biological psychiatry and the medical model of brain disorders had encroached on the province of psychodynamically/psychoanalytically informed psychiatry (Nasrallah, July, 2014, p. 22). In 1987 the Osheroff decision[3] accelerated the trends toward managed health, briefer psychotherapies (cognitive–behavioral therapy, dialectical behavior therapy and interpersonal

10 The succession of psychoanalytic theories

psychotherapy), psychopharmacology, best practice standards and evidence-based practice (McWilliams, 2005). Also, molecular genetics and neuromodulation (brain stimulation) hold promise, respectively, for preventing and treating brain diseases (Nasrallah, July, 2014, p. 49). Now, into this unsettled mix, the latest wave of theoretical innovation emanates from brain imaging and neuroscience research findings. Considering its magnitude and ubiquity, it may be better to refer to this wave as a tsunami.[4]

> In the 1990s, the advent of hi-tech brain imaging techniques … made it possible for neuroscientists to observe, describe, and document with more precision how mammals (humans included) react to a variety of situations. With these developments, findings from neuroscience research (with a lot of initial resistance from the psychoanalytic community) began to infiltrate some psychoanalytical writing with implications for clinical practice. Numerous clinicians and researchers integrated the various domains of discourse … The resistance to applying neuroscience findings to clinical psychological theories is slowly crumbling. But there remains a significant portion of the psychoanalytic community that is either skeptical of or outright rejects the application of neuroscience to clinical practice, arguing that locating the mind in the brain is reductionism. In their view, understanding how the brain functions does not and cannot inform mind. Mind is viewed as constituting meaning, and it is this domain of experience that is, or should be, the only realm of therapeutic understanding and intervention.
>
> (Rustin, 2013, pp. 9–10)

See Summers (2013) for the latest iteration of this meaning-making position.

With respect to integrating neuroscience research findings and empirical research into psychoanalysis, Tyson (2004) suggests that the psychoanalytic community may be getting more accommodating. She proposes a rapprochement between separation-individuation theory and attachment theory held together by "a fully elaborated relational (non-linear, author's emphasis) dynamic systems theory" (p. 507). Questioning the usefulness of both traditional developmental stage theory and the tendency to locate adult psychopathology in earlier developmental epochs, she maintains that contemporary psychoanalysis may be getting ready to embrace the joining of multiple schools of psychoanalysis with different sciences and domains of knowledge. We can see in Tyson's forecast the faint stirrings of theoretical heterogeneity and flexibility presaging the emergence of regulation theory.

From another perspective, Cooper (2008) suggested significant theoretical shifts have taken place, resulting in a possible consensus, based on his informal survey.

> Despite the many individual differences reported in my survey, in the United States a two-person psychology and some version of a comprehensive intersubjective or relational point of view, have superseded traditional resistance analysis as the core theoretical and technical viewpoint for most of the our

profession in the United States. The representational world, conceived of as internalized object relations, has superseded drive-defense configurations as primary determinants of mental life. The intrapsychic focus on the patient has to a great degree given way to or at least been supplemented by scrutiny of the interpersonal – intersubjective – relational transactions between patient and analyst.

(pp. 247–248)

I want to add to this section a note of conciliation amongst this cacophony of perspectives. It seems to me that as the yield from decades of theoretical arguing diminishes, research combining different methodologies might produce sectors of harmony leading to more collaborative study in the search for psychoanalytic "truths."

Two quite different cultures are to be found within psychoanalysis, one more clinical in orientation, more focused on meaning and interpretation, and relying primarily on the traditional case study method, the other more research-oriented, focused on cause-and-effect relationships, and relying primarily on methods borrowed from the natural and social sciences.

(Luyten, Blatt and Corveleyn, 2006, p. 571)

Rather than struggle toward formulating a theoretical integration between the hermeneutic and the neopositivist approaches, the authors advocate methodological pluralism as a way to find fruitful complementarities in these two positions. "Studies using different methodologies, ranging from N=1 studies, multiple case studies, and the study of narratives, to questionnaire research, observational research, and experimental studies, can all potentially contribute to approaching the complexity of psychoanalytic hypotheses and, ultimately, of human nature" (p. 593). Finally,

an exchange between the idiographic level, which aims at assessing, understanding, and treating individuals, and nomothetic research, which aims at discovering 'probabilistic' laws or 'master narratives,' can only lead to better theories, a better understanding of individual patients, and thus a gain for everyone involved.

(p. 595)

Curtis (2008) has attempted to integrate psychoanalysis and the broader field of psychology, viz., clinical and academic psychology. She believes that recent paradigmatic developments in both knowledge domains herald possibilities for change. In psychoanalysis she highlights the shift away from unconscious motivation based on drive to one centered on desire, and the significant challenge to the hegemony of evidence-based, manualized therapies from, for example, Shedler (2010; 2015), who presents compelling accounts of the efficacy of psychoanalysis and psychoanalytically informed psychotherapies. In psychology she presents a growing acceptance of unconscious processes, albeit from a cognitive rather than a dynamic perspective, and the emerging interest in the affective revolution and bottom-up emotional processing. She has formulated the desire-affect model of human motivation grounded in neuroscience and cognition with an emphasis on experience over interpretation as the main force in change, as elements to consider. While her model has much to offer, her proposed changes in terminology and her reformulation of unconscious processes may demand too much from psychoanalysis as it requires contemporary psychoanalysis to move from the theoretical periphery to join psychology in the mainstream center. (These comments were drawn from a review of Curtis' book by Andrew B. Suth, 2011.[5])

I conclude this section with comments on the place of psychoanalysis in the array of disciplines contributing to the contemporary understanding of the human condition. Furthermore, in the following chapters, as the reader becomes familiar with my argument, two central questions need to be addressed. First, does psychoanalysis occupy the core position or is it one among many domains of knowledge and disciplines offering contributions? Second, assuming it is desirable, is it possible to "integrate" these disciplines or can we be content with a synergetic alignment?

Palombo (2017, p. 256) has been giving this problem a good deal of thought as he contemplates the need for a grand psychoanalytic vision. He takes the position that psychoanalysis needs to shift in the direction of scientific realism that integrates various approaches organized by non-linear dynamic systems theory. He includes as components of scientific realism these approaches: intersubjectivity, especially a variant of continental phenomenology (Orange, 1995; 2013 and Stolorow and Atwood, 1983); self psychology, motivational theory, evolution and non-linear dynamic systems theory (Lichtenberg, Lachmann and Fosshage, 2011); and attachment theory (Schore, 1997a; 1997b; Palombo, 2017, p. 256). I take an alternative position, that scientific pluralism offers multiple perspectives, one that is less dogmatic, more welcoming of alternative theories and that is more compatible, or rather comfortable, with postmodern philosophy.

Neuropsychoanalysis

The beginning of the linkage between neuroscience and psychoanalysis may be said to have begun in 1895 with Freud's imaginative *Project for a Scientific Psychology*. It was Freud's hope to create a revolutionary psychology grounded in the emerging discovery of neurons. The *Project* was abandoned because he was unable to fashion a satisfactory neuronal explanation for the three problems that dominated his practice at the time: "the choice of neuroses, the importance of sex, and the mechanism of pathological repression" (Sulloway, 1979, p. 131). This frustration never clouded his life-long hope and search for an integration between the biological and the psychological functional dimensions of the human mental apparatus.

> Had [neuro] imaging been available in 1894, when Freud wrote "On a Scientific Psychology," he might well have directed psychoanalysis along very

different lines, keeping it in close relationship with biology, as he outlined in that essay. In this sense, combining brain imaging with psychotherapy represents top-down investigation of the mind, which continues the scientific program that Freud originally conceived for psychoanalysis.
(Kandel, 2005, p. 386)

The linkage between neurobiology and psychoanalysis goes through the work of Aleksandr Romanovich Luria, a Russian psychologist who is credited with being one of the central figures in introducing psychoanalysis into Russia. Between 1922 to the early 1930s he promoted the psychoanalytic enterprise until the political climate became hostile. He was denounced and "found guilty of ideological deviations" (Kozulin, 1984, p. 20 in Kaplan-Solms and Solms, 2002, p. 29). Fearing for his academic career and even his life, he scrubbed away any reference to psychoanalysis while continuing to identify with and maintain Freud's approach to the study of neuropsychological disorders and the neurobiological bases of behavior, using the clinico-descriptive method. Luria's method of "dynamic localization" involved two stages: qualification of the symptom and syndrome analysis. In the first stage a detailed picture of the symptom's internal psychological structure is obtained in order to elucidate its fundamental psychological basis. The second stage "identifies the *various basic factors* that contribute to the functional system, and at the same time it identifies the *various basic functions* of the different parts of the brain" (p. 41). Luria's contribution facilitates the study of the dynamic nature of more complex processes and "*identifies, from a physical point of view, the elementary component functions out of which each mental structure is constructed*" (Kaplan-Solms and Solms, 2002, pp. 43). In this way, "It could be said that Luria's method is to neurology what Freud's is to psychiatry" (p. 40).

What exactly is neuropsychoanalysis? The term itself, "neuropsychoanalysis," was introduced in 1999 as the title of a journal on the subject published by Karnac Books. The earlier name for the subject was "depth neuropsychology." Interestingly, the term neuropsychoanalysis appears in 1999, the last year of the Decade of the Brain. This decade saw a concerted effort to use computerized brain imaging to map functional aspects of and to discover answers to the mindbrain puzzle. The neuropsychoanalytic enterprise sees itself as the realization of the effort initiated by Freud in the *Project for a Scientific Psychology* to "map the neurological basis of what we have learned in psychoanalysis about the structure and functions of the mind, using neuroscientific methods available to us today" (Solms and Turnbull, 2011, p. 135). The philosophical underpinning rests in Freud's monism, the philosophical belief that he adopted between 1900 until his death in 1939. Monism is "the view that reality is one unitary organic whole with no independent parts" (*Webster's New Collegiate Dictionary*, 1981, p. 737). Because the mind itself is unknowable, it can only be described through indirect means, while externally it can be observed as the brain. So it takes on the dual mindbrain characteristics, subjectively perceiving itself and also objectively visualizing itself. Accordingly, Freud's philosophical position is referred to as "dual-aspect monism." The mind is embodied in the brain and the functionality of both can be studied through indirect inference and direct observation. Freud's quest of describing how the human mental apparatus, from a psychoanalytic perspective, might be represented in brain tissues has been realized. To my knowledge the first such draft hypothesis has been formulated by Kaplan-Solms and Solms' "Towards a Neuroanatomy of the Mental Apparatus," in *Clinical Studies in Neuro-Psychoanalysis* (2002, Chapter Ten, pp. 243–284).[6] It is an impressive undertaking and worth getting acquainted with. Incidentally, Kaplan-Solms and Solms, in addition to the six metapsychological points of view (the economic, the topographical, the dynamic, the genetic, the structural and Erikson's (1950) psychosocial), add a seventh, the physical point of view (Kaplan-Solms and Solms, 2000, p. 251).

If neuropsychoanalysis is understood as the intersection between psychoanalysis and neuroscience, it seems that an interdisciplinary construct is necessary to demonstrate how regulatory processes inform actual clinical work with clients. As headway is made in identifying and correlating physical, anatomical parts of the brain with mind functionality, a different kind of theoretical framework becomes necessary to explain the human condition. One theoretical orientation is simply too small a tent. I propose a developmental algorithm consisting of seven overlapping and complimentary domains of knowledge that combine to form an interlocking explanatory synergy. They are: 1) modern metaphor theory; 2) attachment theory; 3) self psychology with intersubjectivity theory and relational psychoanalysis; 4) cognition; 5) contemporary psychoanalytic developmental psychology; 6) complexity theory; and 7) neurobiology with narrative theory (Bendicsen, 2013, p. 196). The developmental algorithm is an operational explanatory bridge between the abstract (a contemporary psychoanalytic metapsychology) and the practical (the clinical requirements of regulatory processes). In Chapter Three I will present a case formulation using these domains to form a comprehensive explanatory hypothesis.

As I move toward the close of this section on neuropsychoanalysis, consider this amazing paragraph validating the interdisciplinary nature of the clinical process driven by neurobiological research findings. Contemporary psychiatry is studying the efficacy of multiple interventions on brain function. While continuing its emphasis on psychopharmacology, it is recognizing the contributions of psychotherapy and neurobiology to the treatment of mental illness. A proponent of this synergy is the psychiatrist, Henry R. Nasrallah (2013), who goes so far as to suggest renaming the "talking cure" from psychotherapy to neuropsychotherapy.

> Psychotherapy needs to be reconceptualized, rebranded, and repositioned as a neurobiological treatment – because, in fact, that's what it is. This notion goes hand-in-hand with unimpeachable evidence that *the mind is an integral component of the brain and mental illness is generated from genetic or environmentally induced dysregulation of neurobiological homeostasis.*
>
> ...
>
> An important line of evidence for the neurologic effects of psychotherapy are studies of positron-emission tomography showing that psychotherapy

induces changes in specific brain regions that are identical to changes induced by drug therapy. The component activities of psychotherapy – verbal and nonverbal communication, evocation of memories, empathizing, challenging, connecting the dots, triggering insights, and reducing anguish – are transduced into instantaneous neuroplastic changes, which can be lasting and lead to corrective modification of the neural circuitry of feelings, thinking, and behavior.

...

Most non-neuroscientists might not be aware that the brain changes continuously, moment to moment, forming dendritic spines that immediately encode verbal and nonverbal in response to experiences throughout life. A skilled psychotherapist exploits this biological property of the brain to relieve the anguish and psychopathology of its avatar, the mind.

(Nasrallah, 2014, p. 19)

Anticipating a neuropsychoanalytic metapsychology

Freud's "Papers on Metapsychology" (1915) continued his thinking on creating a grand psychological theory that began with the "Project for a Scientific Psychology" (1895) and extended through the well-known seventh chapter of the *Interpretation of Dreams* (1900) (Freud, 1915, pp. 105–107). Freud wanted his psychology to be accepted by the scientific establishment and to do so grounded his conceptual framework in the dominant hydraulic-energy-drive/instinct-evolution paradigms of his day. To distinguish his clinical psychological observations, obtained through empirical data gathering, from theoretical speculations, he used the term metapsychology (meta = beyond) to denote a higher level of abstraction in explaining his metaphor laced paradigm. Freud's elegant model, over 100 years old, has demonstrated remarkable adhesive power (Prochnik, August, 14, 2017). However, it is considered anachronistic judging by the advancements in science today. Challenges specifically to Freud's "Energy-drive" model, and to psychoanalysis in general, are well known, formidable and growing (Kernberg, 2011; Schechter, 2014). Theorists struggle to reconfigure the once monumental model as bridging concepts are sought to establish explanatory compatibility with diverse knowledge domains such as biological psychiatry, academic psychology, non-linear dynamic systems theory and neurobiology, to name a few. Can psychoanalysis survive as a stand-alone discipline? Where are the emerging mindbrain metapsychologies?

Let us examine one of the most recent. Imbasciati (2017) has advanced his protomental theory, a framework consistent with neuroscience.

My theory ... is a hypothesis (as was Freud's one) which attempts to connect the discoveries and concepts of the different sciences of the mind in a general explanatory theory of the origins and functioning of the mind, so that by drawing on a comparison with the various sciences, it can be useful for psychoanalysis: for the clinical practice of psychoanalysis and above all for the public image of psychoanalysis in the general scientific panorama.

(Imbasciati, 2017, p. 150)

Standing in the way of a neuropsychoanalytic metapsychology is the adherence, on the part of the psychoanalytic establishment, to the "Energy-drive" model. The distinction between "empirical discovery" and metaphoric "instrumental invention" with respect to key concepts such as repression, drive and libido is raised as a critical epistemological question. Discovery should not be confused with theory: "theory is the terminal of a series of hypotheses connected organically between one another, invented to give better comprehensible form to what has been observed" (Imbasciati, 2017, p. 196). And "A theory is a conceptual invention, an instrument to understand new discoveries, to refine the method, to make a science progress: an instrument which is therefore provisional, as well as hypothetical" (Imbasciati, 2017, p. 202). Core concepts in his metapsychological theory include:

1) the mind and brain are a functional unity with reciprocating influences;
2) the vast majority (95%) of our mindbrain works beyond consciousness;
3) while genetic expressivity/biological heritage is the same for all homo sapiens, epigenetic variability/individual experience allows for an infinite possibility of differentiation; in other words, while the macro-morphology of the human brain is determined by the genome, its micro-morphology, and specifically its physiology, are structured according to an individual processing function determined by the experience of each individual subject;
4) all human mindbrains are different; there is no normal mindbrain, but its mental expressions and behavior enable us to speak of what appears as average for all adult human beings;
5) experience is understood idiosyncratically and never corresponds exactly to external reality;
6) in the experience that has structured the mindbrain, the most comes from the emotional information derived from inter-human relations;
7) the structuring begins in the fetus with its greatest impact in the first eighteen months and continues throughout life; given a suitable emotional level, the mindbrain is in a constant state of being restructured;
8) every single detail of all mental life and behavior is regulated by the mindbrain including temperament, character, affect and cognition;
9) what is scientifically called "mind" today does not correspond to what the individual believes is happening inside him, or what he believes is happening or what he is conscious of;
10) what are called emotions do not coincide with what we are feeling; most of the work done by the mindbrain is emotional and, at the biochemical level, is beyond our awareness;

11) all of the right hemisphere and most of the left hemisphere constitute the emotional brain; the processing of the emotional brain is transformed, in circuits in the left brain, into cognitive operational skills which are possibly conscious; so that it can be said that a human being "reasons" on the basis of the emotional work of the mindbrain.
12) in the first eighteen months the mindbrain is operationalized and regulated by emotions which are encoded into functional, implicit memory, not memory of contents which can possibly be re-evoked.
13) the mindbrain's work depends on the interconnectivity of billions of neurons which are constructed by experience; each experience constructs new neural connections or assemblies; the genome determines the number of neurons while experience generates their connections, called the "connectome."
14) memory also depends on the connectome; what is remembered may not correspond to what is memorized. Memory is a continuous, ever-changing activity, not a warehouse. Even though memory regulates all our abilities, very little of it is remembered.

(Imbasciati, 2017, pp. 61–63)

It is apparent that Imbasciati is not attempting to fashion an integration among the sciences with psychoanalysis. The question of understanding the origins and functioning of the mindbrain will, in all likelihood, rest on collaborative efforts organized according to a pluralistic approach (see Chapter Four in this monograph). With influential thinkers and organizational leaders abandoning psychoanalysis' one-hundred-year-old anachronistic energy-drive model, its contribution to today's rapidly expanding knowledge base is expanding. Psychoanalysis urgently needs a new, contemporary metapsychology. When that day arrives, psychoanalysis will reclaim its relevance and its vital role at the multi-theoretical table.

Another attempt at a neurobiologically informed metapsychology comes from Sripada and Jobe, entitled "A Biological Topography for Psychoanalysis" (October 25, 2017, unpublished). Their project will "Consider whether brain structure functioning, and its intermingling of visceral, limbic, and cortical domains of conscious experience, offer clearer psychoanalytic explanations than some structural topographic concepts bases on Id, Ego, and Superego." Seeking to advance the integration of the natural sciences with psychoanalysis, Biological Topography does not rely on the rider and the horse metaphor of id, ego and superego, but rather the structures and functions of the body. Mark Solms made significant contributions toward the integration of the natural sciences and psychoanalysis, but he relied on the id, ego and superego metaphor and located core consciousness within the brain stem. Sripada and Jobe take the whole body into account (Personal communication with Dr. Sripada on April 2, 2018). An anticipated publication of their work will explicate their views.

Let us now turn to a subject which will play a central role at the theoretical table, that of regulation theory.

Notes

1 This paper won the first Yellowbrick Foundation Emerging Adult Paper Prize on June 20, 2014 sponsored jointly by The Yellowbrick Foundation and Chicago Psychoanalytic Institute (Yellowbrick, 1560 Sherman Avenue, Suite 400, Evanston, IL 60201 and the CIP, 122 South Michigan Avenue, Suite 1300, Chicago, IL 60603).

   An abridged version of this paper was published in the *Illinois Society for Clinical Social Work Newsletter*, Spring, 2014 under the title of "The Argument for a Compatible Set of Interlocking Theories as Applied to a Case Example" (ISCSW, P.O. Box 2929, Chicago, IL 60690–2929).

   An abridged, updated version of this paper entitled, *The Regulation Hypothesis: A Framework for Focusing on the Self in Treatment with Emerging Adults*, was published in the *Yellowbrick Journal of Emerging Adulthood*, Issue V, Tenth Anniversary Issue, December, 2016, Yellowbrick Foundation, 1560 Sherman Avenue, Suite 400, Evanston, IL 60201.
2 Freud's response to the threat to theoretical homogeneity has many historical precedents. One I am reminded of in this regard is that of Pope Paul III's response to the Protestant Reformation. In 1540 the Pope enlisted Ignatius Loyola and his Society of Jesus to define Roman Catholic orthodoxy through education to the public and spearhead the Counter Reformation. In 1542, the Pope's education initiative was followed by an obedience initiative, "The Sacred Congregation of the Roman and Universal Inquisition or Holy Office," the Roman Inquisition (Catholic Encyclopedia).
3 In 1979, Dr. Rafael J. Osheroff, a 42-year-old white male, married with three children, a medical doctor with a specialization in nephrology, was admitted to Chestnut Lodge hospital in Maryland with a diagnosis of depression and narcissistic personality disorder. Psychoanalysis was the only treatment offered. At the seven-month mark in treatment, his condition worsened to the point where his parents intervened and transferred him to Silver Hill Foundation hospital in Connecticut. At Silver Hill he was diagnosed with psychotic depression and placed on psychotropic medication, including phenothiazines and tricyclic antidepressants. Improvement was dramatic and within three months the patient was back at work. Dr. Osheroff initiated a right to effective treatment law suit which was ruled in Dr. Osheroff's favor in an out of court settlement in 1987 (Klerman, 1990; Packer, 2012).
4 The contemporary collaboration between neuroscience and psychoanalysis got off to a bad start. Building on the discovery of REM sleep by Aserinsky and Kleitman in 1953 and the demonstration that REM sleep was the external manifestation of the subjective dream state (Aserinsky and Kleitman, 1955 and Dement and Kleitmann, 1957a, 1957b), neuroscientific research by 1975 had produced a detailed picture of the anatomy and physiology of "dreaming sleep."

   > This picture, which is embedded in the *reciprocal interaction* and *activation-synthesis* models of McCarley and Hobson (1975, 1977), has dominated the field ever since – or at least, as we shall see, until very recently. These authoritative models proposed that REM sleep and dreaming were literally "switched on" by a small group of cells situated deep within the pons, which excreted a chemical called acetycholine. This chemical activates the higher parts of the brain, which are thereby prompted to generate (meaningless) conscious images. These meaningless images are nothing more than the higher brain making "the best of a bad job … from the noisy signals sent up from the brain stem."
   >
   > (McCarley and Hobson, 1977, p. 1347)

   After a few minutes of REM activity, the cholinergic activation arising from the brainstem is counteracted by another group of cells, also situated in the pons, which excrete two other chemicals: noradrenaline and serotonin. These chemicals "switch off" the cholinergic activation and thereby, according to the theory, the conscious experience of dreaming (Solms, 2015, p. 129).

> Thus all the complex mental processes that Freud elucidated in the *Interpretation of Dreams* (1900) were swept aside and replaced by a simple oscillatory mechanism by means of which consciousness is automatically switched on and off at approximately 90 minute intervals throughout sleep by reciprocally interacting chemicals that are excreted in an elementary part of the brain that has nothing to do with complex mental functions. Thus, even the most basic claims of Freud's theory no longer seemed tenable.
>
> (Solms, 2015, p. 129)

With considerable certitude it was announced that

> The primary motivating force of dreaming is not psychological but physiological since the time of occurrence and duration of dreaming sleep are quite constant suggesting a pre-programmed, neutrally determined genesis. In fact, the neural mechanisms involved can be precisely specified … if we assume that the physiological substrate of consciousness is in the forebrain, these facts [i.e., that REM is automatically generated by brainstem mechanisms] completely eliminate any possible contribution of ideas (or their mental substrate) to the primary driving force of the dream process. [Hobson and McCarley, 1977, pp. 1346, 1338].
>
> (Solms, 2015, pp. 129–130)

That is where matters stood until "a second body of evidence gradually began to accumulate, which led some neuroscientists to recognize that *perhaps REM sleep was not the physiological equivalent of dreaming after all*" (Solms, 2000, in Solms, 2015, p. 130). The basis for assuming that dreaming was a feature of REM sleep was based on the statements from awakened dreamers in REM sleep who reported that in 70–95% of cases that they were dreaming, while those in non-REM sleep reported that they were dreaming in only 5–10% of cases. An isomorphism slowly emerged to suggest that, whatever the strong correlation between REM sleep and dreaming, that dreaming was caused by REM sleep. In other words, randomized REM stimuli rendered dreaming devoid of deterministic meaning.

Unexpected evidence has emerged to the effect that REM sleep and dreaming are caused by different brain mechanisms. The hypothesis that two separate brain mechanisms exist, one for REM sleep and another for dreaming, can be easily tested by using the clinic-anatomical method. In this method known parts of the brain are significantly damaged or obliterated (in humans through traumatic accident or disease), and if REM sleep or dreaming is eliminated, then it can be assumed that the damaged part is responsible for the loss in specific functionality. We now know that damage to the pons (and nowhere else) leads to a cessation of REM sleep in lower mammals. In the neurological literature in twenty-six cases with damage to the pons, REM sleep was eliminated, while dreaming continued in twenty-five of twenty-six cases.

> The parts of the brain that are crucial for dreaming and those that are crucial for REM sleep are widely separated, both anatomically and functionally. The parts of the brain that are crucial for REM are in the pons, which is located in the brainstem, near the nape of the neck. The parts of the brain that are crucial for dreaming are, by contrast, situated exclusively in the higher parts of the brain, in two specific locations within the cerebral hemispheres themselves.
>
> (Solms, 2015, p. 133)

"The first of these locations are deep in the white matter of the frontal lobes just above the eyes. The main function of this higher pathway is to motivate the subject to seek out and engage with external objects that can satisfy its inner biological needs." This is Panksepp's SEEKING system, a primary motivator of human behavior (Panksepp, 1985, p. 273 in Solms, 2015, p. 133–134). (See also Chapter Five of this book.) Panksepp assigns the SEEKING system responsibility for "foraging, exploration, investigation, curiosity, interest [and] expectancy" (Panksepp, 1998, p. 145 in Levin, 2015, p. 136). It is believed now that dreaming can be switched on and off by a neurochemical oscillator involving dopamine.

> The second location is a portion of the grey cortex at the back of the brain (just behind and in above the ears) called the occipito-temporo-parietal junction. This part of the brain performs the highest levels of processing of perceptual information and is essential for the conversion of concrete perception into abstract thinking.
>
> (Solms, 2015, p. 136)

Levin sums up this groundbreaking neuropsychoanalytic discovery:

> In the past, there was a debate as to whether dreams were emotionally meaningful and to what extent they are a byproduct of neuronal firings in the brain. However, now there is a consensus that the debate has been settled in favor of the assumption of psychoanalysis that dreams are meaningful.
>
> (Levin, 2015, p. 132)

5 Reflecting on Curtis' attempt to integrate psychoanalysis and psychology, I find that Bucci's (2000, in Curtis, 2008) paper helps to expand the argument for integration. Bucci advocates for a "psychoanalytic psychology" in the cognitive sciences that corrects the failure of cognitive psychology to acknowledge its debt to psychoanalysis. She contends that cognitive psychology has already incorporated the following concepts: "mental models, mind-body interaction, unconscious processes, dual processes of thought, and naturalistic research milieus" (p. 203). The fact that the various subfields of psychology (excluding psychoanalytic psychology) are taught in universities while psychoanalysis is taught in free-standing institutes, generally away from scientific scrutiny, contributes to the isolation.

6 Hadley (1989) has anticipated Solms' (2002) outline of the neuroanatomy of the mental apparatus by defining the neurobiological substrate of Lichtenberg's (1989) motivational systems. Hadley states that all psychological functions are whole brain involvements with given functions involved in certain areas more prominently than others. There are two distinct organizing principles underlying motivational mechanisms. First is the maintenance of familiarity of neural firing patterns; this of course refers to the Hebb rule – neurons that fire together, wire together. "This process depends on familiarity of stimuli (both internal and external) and is measured by comparator mechanisms in the limbic system, notably the hippocampal comparator mechanism (Deadwyler, 1987). It is the neurobiological substrate for the repetition compulsion."

"The second process supporting motivation is the pleasure-punishment principle, which supplies both positive and negative motivations depending on the outcomes of the matching process and the addition of affects. This process is the basis for the pleasure principle."

"These two processes are mediated through the basic information processing equipment of the limbic system and subserve all motivational subsystems" (Hadley, 1989, p. 337). "The execution of appropriate behavioral responses indicated by the results of information comparison is mediated through hypothalamic and basal ganglia activation, which have been preprogrammed either innately or by learning" (p. 338).

Hadley's project matched Lichtenberg's five motivational systems to anatomical structures and their functionality in an impressive display of integrating neurobiological processes with psychic regulation. While these systems are functionally distinct, they link up with each other through engagement and disengagement processes allowing for enormous variety, complexity and flexibility in response to stimuli. "However, it also permits distortion of motivation and spillover from one system to the other. An example can be observed in eating disorders, where the physiological motivations are used in the service of control and effectance" (p. 338).

2

REGULATION THEORY

Toward a new paradigm

As I move through various training and supervisory experiences with budding clinicians, it is abundantly clear that, for some, their overarching concern is expressed anxiously as "I don't know how all the (theoretical) pieces fit together!" We have accumulated a kaleidoscopic array of theories grounded in metapsychology (the dynamic, economic, topographical, psychogenetic, structural, adaptive, psychosocial and now physical models), those grouped as experience-distant or experience-near theories, as well as those considered as one-, two- and three-person psychologies. In addition, most of the authors of these theories have contributed a significant autobiographical component, so that to fully understand these frameworks an appreciation of their historical and cultural contexts is necessary (Bendicsen, 2013). While Noy (1977) and Pine (1990, pp. 22–23) may consider such diversity a strength, that opinion does not seem to be shared by many students and practitioners. See Mitchell and Black (1995) for a discussion of an attempt at managing this conceptual diversity in the emerging field of "comparative psychoanalysis" (p. 207).

Not infrequently, the clinician's initial attempt at integration is a kind of default action in that the automatic setting is "eclecticism." The eclectic setting, a form of reductionism, is to cherry-pick the best of the theories as one understands them at a point in time to fit a particular diagnostic/treatment case need (Skurky, 1990 in Palombo, 2013). Usually the result is constantly changing one's theoretical amalgam. The end result of this theoretical customization is that the novitiate, who struggles to come to grips with this massive metaphorical stew, is confused, resulting in conceptual incoherence; consequently, the patient/client is ill served. It is in the application of these theories to case formulation and treatment planning that the struggle for coherence begins in earnest. No wonder when asked, "What is your theoretical orientation?" clinical students have a "deer in the headlights" look, a look that applies to the novice and many graduates alike. See Chapter Four in this monograph for elaboration.

Schafer (1979) has given this problem some thought. He elaborates on the development of the professional identity of a psychoanalyst, in particular the many tensions associated with acquiring and maintaining a theoretical orientation. He begins with the assertion that the analyst is always in the state of becoming an analyst. As stations are reached in this endless journey, specific tensions need to be addressed. The first tension occurs as a result of the school within which the analyst trained. A relevant set of ideas has been transmitted which can acquire the status of immutable truths in the context of closed thought systems. When "truths" from another school are encountered, tension is inevitable in attempting reconciliation. The second tension results from the sense of confinement from working in a specific theoretical school. Challenging conceptual truths the analyst has used as a reliable anchor can lead one to have loyalty and identity conflicts because these truths were taught by idealized teachers and supervisions. The third source of tension emanates from one's success in understanding the patient's change factors. Improvement may result from forces outside of therapy or from techniques informed by and associated with alternative schools. The fourth tension arises from how much emphasis to place on time spent on inner fantasy versus the outer reality of everyday experience. The fifth and last tension Schafer discusses is the need to retain an affirmative attitude toward the patient's productions, regardless of how troubled the patient is, lest the empathic stance be undermined.

Because one theory does not have the overarching explanatory power that a *compatible set of interlocking theories* might, the search becomes one for theoretical alignment, or goodness of fit, amongst frameworks. In this article I will argue that regulation theory offers a solution. Incidentally, Palombo (2017) also has been working on the problem "of bringing some kind of rapprochement among these theories" to embrace a different model of the development of the self, one informed by complexity theory. See Chapter Five for a discussion of this issue.

Charles Jaffe (2000) captured the problem well:

> The study of adolescents strikes me as a bustling marketplace of issues. Picture the open markets in Florence during the Renaissance or London's Portobello Road of today. In the marketplace of adolescent psychology, everyone gathered seems to have one thing in common: that somehow, roughly in the second decade of life, people transform from a state in which they mostly behave and think like children to a state in which they mostly do not. Merchants from diverse cultures converge with a variety of wares in the market place. They loudly trumpet their products and proclaim their power and efficiency to understand development, to describe its problems, to supply a fix. Biologists abound. They proclaim the power of their hormones to jump-start development and to shape body and brain. Psychoanalysts, having placed many of their kiosks close together, intend to drown out the cries of competitors with claims of superior power to explain and effect change. In this niche of the market especially, merchants tend to want their wares to dominate and others' to occupy a subsidiary role at most. Academics fill every spare inch.

> They act as monitors of truth in advertising, as referees reminding others not to occupy more space than they deserve, and as perspective enforcers who remind merchants that their touted products are merely necessities of modern culture, with questionable enduring value or universal marketability.
>
> (pp. 40–41)

Most learning takes place against resistances. One of the most powerful of resistances is the tendency of a group to defend itself against perceived corruptions to its entrenched belief system. Psychoanalysis has been compared to religious fundamentalism.

> Just as fundamentalists regard their system as infallibly and literally accurate, members of psychoanalytic movements adhere to the doctrine of their theory. Although at times members of any part of a group may add ideas to the theory, the structure of the doctrine is beyond question.
>
> (Summers, 2006, p. 8)

With the hegemony of ego psychology now history, the time for a single overarching explanatory framework of psychological understanding emerging at this time seems remote indeed. While some may continue searching for a *Weltanschauung* (Freud, 1933, p. 158), it seems to me that a contemporary explanatory synergy will need to emerge with various domains of knowledge contributing elements of understanding. The likeliest elements in a *compatible set of interlocking theories* will include: 1) modern metaphor theory; 2) attachment theory; 3) self psychology with intersubjectivity theory and relational psychoanalysis; 4) cognition; 5) contemporary psychoanalytic developmental psychology; 6) non-linear dynamic systems theory (or complexity theory); and 7) neurobiology with narrative theory. I maintain that these seven elements will become part of a contemporary explanatory system known as regulation theory (Hill, 2010). It will draw from both empirical/neopositivist and social constructivist philosophies. An example of this synergy can be found in Bendicsen's (2013) *The Transformational Self: Attachment and the End of the Adolescent Phase* (London, UK: Karnac Press).

From this point it is necessary to prepare the reader for a shift in terminology. It will be helpful to have a basic knowledge of brain anatomy, which can be obtained easily by consulting the brain diagrams in, for example, the Cozolino (2006; 2010) books or related sources. Constraints on length do not allow me to define or otherwise elaborate on each piece of terminology.

Two perspectives on brain organization – the triune brain and the social brain

Let us consider the distinctions between the two major perspectives on brain development and organization as a pathway to the subject of neurobiological regulation. The triune brain hypothesis was formulated by Paul MacLean (1990). It emphasizes brain development from an evolutionary framework, similar to the models of Darwin and Freud. The oldest part, the core or the brain stem, is equivalent to the reptilian brain. Its functions include activation, arousal, homeostasis, reproductive drives and primitive memory. The next to evolve, the limbic system or the mammalian brain, wraps around the core and is the component for emotionality. It is central to learning, memory and social interaction. The last part to evolve is the cerebral cortex, responsible for conscious thought, problem solving, language and self-awareness. These three systems are organized essentially on the basis of the age of each element or layer and the architectural proximity of system elements to each other. MacLean believed that the three brains constitute different mentalities and consequently do not communicate or work well together. The inadequate integration of these three cohabiting brains gives rise to the metaphor of imagining simultaneously treating a crocodile, a horse and a human (Cozolino, 2010, pp. 5–6). As useful as the triune brain is, it does not comport well with the recent findings from neuroscience research.

> Although MacLean's model is a helpful place to begin, it soon becomes inadequate in explaining many of the brain's complexities. In reality, MacLean's reptilian and paleomammalian brains have continued to evolve along with the neomammalian brain. Newer systems, emerging to address the changing requirements of survival, have conserved, modified, and expanded components of preexisting systems. Another problem with MacLean's model is that there are no clear delineations between layers, with regions such as the insula and cingulated cortex possessing legitimate "dual citizenship" of both the paleo- and neomammalian cortices.
> Furthermore, all three layers are linked together in complex vertical neural networks, thereby allowing the whole brain to coordinate everything from simple motor movements to complex abstract functions.
> (Cozolino, 2006, p. 25)

The social brain hypothesis (Brothers, 1990; Barton and Dunbar, 1997) is based squarely on the functionality of system components. The social brain hypothesis (SBH) originated as an alternative explanation to the idea that brains evolved in size to process ever more complex information of ecological relevance, such as problem solving. Additional evolutionary anthropological research suggests that social group size can be correlated with cortex size in that the complexity of maintaining social interactional stability makes enormous cognitive energy demands requiring a larger cerebral cortex (Cozolino, 2006, p. 12). According to Moll and Tomasello (2007), while the theory of social competition may have driven the evolution of primate intelligence, it is social cooperation that has enabled humans alone to develop language, culture and civilization. In other words, human social interaction is so vast and differentiated that additional executive brain mass is required to accommodate this increased specialized activity (Dunbar, 1992, 1993, in Cozolino, 2006). The SBH, therefore, offers more explanatory usefulness than the triune brain when considering questions of relatedness, attachment and pathology.

The shift from the preoccupation with the evolutionary/structural brain to its more pertinent functionality requires a significant reorientation from the traditional perspective in three ways. First, the cortical lobes expand from four, including the parietal, temporal, occipital and prefrontal, to six with the addition of the cingulate and insula. While the first four are located on the surface of the brain, the last two are buried deep in the brain's interior. Second, cerebral dominance for social and emotional functioning is the right hemisphere, not the left. Third, the brain is not regarded as fully formed at any point in its development, but rather is in a constant process of dynamic formation and reformation, construction and reconstruction, across its entire lifespan (Cozolino, 2006, p. 50).

The social brain consists of seven cortical and subcortical structures, all intimately connected to the limbic system. Following Cozolino's outline (2006, pp. 51–65) let us briefly review each of these structures. First, the orbital medial prefrontal cortex (OMPC) is located at "the apex of the neural networks of the social brain" (p. 54). The OMPC is a dual citizen of the cerebral cortex and the limbic system. It serves as the junction for internal and external sensory systems. Its direct connections to the hypothalamus allow it to integrate information from the sensory systems. It serves an inhibitory role in autonomic functions, enabling it to play a major part, along with the right hemisphere, in affect regulation and attachment. The OMPC's vast interconnectedness allows it to link the sensory systems with social information to guide perception, actions and for emotional and sensory information.

Second, the somatosensory cortex is located in the anterior portion of the parietal lobes. It, along with the insula and cingulate cortices, is responsible for containing multiple body representations through our experience with touch, temperature, pain, joint position and our visceral state. The resulting "somatic self" is from birth experience dependent and, as such, is our model for attachment to others.

Third, the cingulate cortex is a primitive cluster of neural activity for somatic, autonomic and emotional information beginning very early in life. It provides vital connectivity for facilitating the neural infrastructure for maternal behavior, social cooperativeness and empathy.

Fourth, the insula cortex is situated deep within the brain and its functions are related to internal experience. Its extensive interconnectivity has suggested its description as the "limbic integration cortex" (p. 56). It allows us to be connected to what is happening to our bodies and so facilitates what might be called "inside-outside congruence" (from W. Gieseke, private correspondence, October 3, 2013). Facial expression, eye gaze and the appraisal of untrustworthiness in others activates the insula. The insula cortex is involved in mediating the entire range of emotions.[1]

Fifth, the amygdalae, with its intimate connectivity to the autonomic system, specializes in the appraisal of danger, mediates the flight-fight response and emotional memory. "The primary role of the amygdalae in the social brain is to modulate vigilance and attention in order to gather information, remember emotionally salient events and individuals, and prepare for action." The amygdala networks become vital in the recognition of faces, facial expression, especially those faces judged to be untrustworthy and threatening (Cozolino, 2006, p. 167).

Sixth, the hippocampus is closely linked to the parietal lobes which aid the hippocampus in complex visual-spatial processing. The hippocampus and its connectivity to adjacent structures are responsible for the organization of spatial, sequential and emotional learning and memory. It is more associated with the processes of the right hemisphere than those of the left hemisphere.

Seventh, the hypothalamus translates conscious experience into bodily processes. Through its regulation of temperature, hunger, thirst and activity levels it is closely tied to the early experience of building body and brain. Through the production of specific hormones the hypothalamus regulates sexual behavior and aggression.

Neurobiological regulation

A cursory examination of the subject "regulatory theory or regulatory systems" reveals many definitions from many disciplines. I will consider five regulatory models and then attempt to formulate a beginning theoretical integration of regulation. The first two are from Hofer (1995; 2006) and Schore (1994; 2003a) and are significant in that the authors attempt to tie their forms of regulation to contemporary psychoanalysis. The third is from Siegel (1999), the fourth is from Cozolino (2006) and the fifth is from Harter (2012). I will only highlight the first three and elaborate in more detail on Cozolino's and Harter's models.

Hofer's (1995; 2006) laboratory research on infant animals led him and his colleagues to reformulate his assumptions about separation and the "nature of the child's tie to the mother" (p. 203). Hofer discovered a set of hidden regulators including "multiple sensorimotor, thermal-metabolic and nutrient based events" that exerted unexpected regulatory effects on mother-infant dyad. Hofer defined the term regulation to "convey the sense that individual systems of the infant were found to be controlled in their level, rate, or rhythm by particular components of the ongoing social interaction with the mother" (p. 203).

Schore's extensive neuroscience literary research (1994) enabled him to develop the groundbreaking hypothesis "that attachment theory is fundamentally a regulatory theory; that is, *the primary function of attachment is that of regulating the child's affect states.*" This regulatory activity emerges early in the right hemisphere and is modulated by mutual facial expression between mother and infant through the language of affect exchange (Palombo, Bendicsen and Koch, 2009, p. 325).

Siegel (1999) privileges emotions as the unifying dimension in self-regulation, agreeing with Luc Ciompi (1991) that

> emotions function as "central organizers and integrators" in linking several domains: providing all incoming stimuli with a specific meaning and motivational direction; connecting mental processes "synchronically" and "diachronically" (within one time and across time); creating more complex interconnections among abstract representational processes that share emotional meaning; and simultaneously attuning the whole organism to the current situational demands on the basis of past experience through

neuro-physiologically mediated peripheral effects. Such organizing features intimately link what are traditionally considered the mental, social, and biological domains. As Alan Sroufe has pointed out, then, emotions are inherently integrative in their function.

(Sroufe, 1996 in Siegel, 1999, pp. 239–240)

Cozolino's model is composed of four distinct regulatory processes that comprise the social brain: the stress regulation system, the fear regulation system, the social engagement system and the social motivational system. I will elaborate on these regulatory systems in the following pages.

Harter's framework is tied to the attainment of regulation in late adolescence based on the maturation of the prefrontal cortex and its emphasis with connectivity with the left hemisphere. This allows for enhanced executive functioning and the implied intersubjectivity dynamics of mutual recognition governing the need to maintain positive interpersonal relationships. Harter's model will function as a bridging concept, allowing me to demonstrate the essential role of the Transformational Self.

A suggested epigenetic framework for regulatory processes

Is there a way to tie together these five perspectives on neurobiological regulation? I pondered this question in my book, *The Transformational Self: Attachment and the End of the Adolescent Phase* (Bendicsen, 2013), suggesting that the varied perspectives on regulation were in need of a unifying framework. Siegel (1999) suggests fitting some of these regulatory dimensions into a pathway using an epigenetic developmental framework (p. 274). Let me extend his framework into a five-part stage process recognizing the obvious overlap and extensive interconnectivity amongst regulatory forces: 1) Hofer's "hidden regulators" may be thought of as operational in utero and immediately after birth forming the scaffolding for internal homeostasis; 2) Schore's "attachment as affect regulation" evolves with the experience of the infant-mother dyadic attunement process and sets the stage for the wider world of social interaction. "As infancy gives way to the toddler period, dyadic regulation is supplanted by 'caregiver-guided self-regulation,' in which the adult helps the child begin to regulate states of mind autonomously" (Siegel, 1999, p. 240); 3) Siegel's emphasis on emotional self-regulation seems especially relevant in the elementary school years as peer interaction and school protocols contribute to shape learning motivation and integrative brain functionality; 4) this phase is followed by Cozolino's "stress regulation system," the HPA (hypothalamic-pituitary-adrenal axis) complex of hormone regulation, which obviously drives puberty (Sklansky, 1991). It is one dimension of the four-part social brain regulation framework that, as an overlay, can be understood as a biological regulation blueprint for the entire life span; 5) Harter's two-part "adolescent self-regulation" position forms as a result of the transition into young adulthood. I end with Luc Ciompi/Seigel's position that emotions, and their developmental differentiation, guide self-regulation. This position may be said to unify self-regulation as both an epigenetic process with distinct developmental markers, as well as a process for the full life cycle.

For the sake of continuity, I will continue to follow Cozolino's outline on social brain regulation (2006). The two categories of regulation I will focus upon are those from neurobiological regulation and emerging self-state/state of mind regulation.

Types of regulatory systems in the social brain

Let us first consider neurobiological regulation. Cozolino (2006, pp. 59–62) outlines four types of regulatory systems found in the social brain. The human body must employ inherent self-righting and continuous stabilizing properties in order to retain cohesion and coherence. Involved in this effort are "the maintenance of internal homeostatic processes, balancing approach and avoidance, excitation and inhibition, and flight and fight responses" and the control of "metabolism, arousal and our immunological functioning. These regulatory processes control our own as well as others' emotional and biological states" (Cozolino, 2006, p. 59). I will lay out his framework.

1) *The stress regulation system is the HPA complex of hormone regulation.* The HPA system regulates the secretion of hormones involved with the body's response to stress and threat. Quick reaction to a threat is necessary for short-term survival while long-term survival depends on the return to normalization after the threat has passed. Prolonged stress can result in damage to the system and even breakdown. The long-term effects of trauma, attachment disturbances, abuse and deprivation are mediated by the HPA system.

2) *The fear regulation system is the orbital medial prefrontal cortex (OMPFC) – amygdala balance.* The fear regulation system is initiated by the amygdala triggering a fight or flight response which then activates the sympathetic branch (SNS) of the autonomic nervous system (ANS). This system becomes operational immediately, generally operates outside of conscious awareness and manifests through anxiety, agitation and/or panic states. So the amygdala is responsible for pairing stimuli with a fear response to enhance prospects for survival and facilitate adaptation. The counterpart to the amygdala/SNS partnership is the OMPFC which can inhibit the amygdala based on conscious awareness. The OMPFC assesses the safety-danger status in the environment, including attachment schema. Through repeated experiences the fear regulation system learns to reset itself, awaiting the next event.

3) *The social engagement system is the vagal system of autonomic regulation.* The tenth cranial nerve is called the vagus (L. for wandering) because it is not one nerve fiber, but rather is a complex network of nerves communicating among the spinal column, heart, lungs, digestive tract, eyes and ears and throat. Its extensive sensory/afferent (conducting inward) and motor/efferent (conducting outward) fiber connectivity contain an efficient feedback loop linking the brain and body components "to promote homeostatic regulation and optimal maintenance of physical health and emotional well being." A vital part of the ANS, the vagal nerve complex works to enhance digestion, growth and social communication through its extensive

connectivity with facial musculature. When stimuli associated with challenge/ danger arises, a decrease in vagal activation, called the vagal brake, facilitates SNS arousal, high energy output and the fight/flight/freeze response.

More specifically, Porges (2001; 2009 in Quillman, 2012) postulates an ANS not organized around traditional alert/awake vs sleep/rest balance, but rather hierarchically (top-down and bottom-up) with the polyvagal parasympathetic branch (PNS) as a core component. The polyvagal network constitutes a social engagement system in that when people feel safe it deactivates the SNS, allowing a nominal level of anxiety without having to disengage socially.

> Social engagement is negotiated primarily via voice tone (prosody) and the striated muscles innervating the face and inner ear. Because this system is myelinated, it allows for very rapid adjustments in the facial muscles and the larynx, permitting people to use shifts in facial expression and voice tone to communicate.
>
> (Quillman, 2012, p. 2)

When danger is perceived, the so-called vagal brake is activated, the PNS goes off line and the SNS is mobilized for action. If the danger is extreme and overwhelming the ANS may shift into dorsal vagal shutdown (a hypoarousal state), switch to the slower unmyelinated vagus nerve, a condition known clinically as dissociation.

4) *The social motivation system includes reward representation and reinforcement.* "Nelson and Panksepp (1998) postulate the existence of a 'social motivation' system modulated by oxytocin, vasopressin, endogenous endorphins, and other neurochemicals related to reward, decreased physical pain, and feelings of well-being" (Cozolino, 2006, pp. 61–62). Derived from more primitive approach-avoidance and pain regulation circuitry, the social motivation system extends into the amygdala, the anterior cingulate and the OMPFC. Fisher (1998) divides the social motivation system into three categories: 1) bonding and attachment (regulated by peptides, vasopressin and oxytocin); 2) attraction (regulated by dopamine and other catecholamines); and 3) sex drive (regulated by androgens and estrogens). The ventral striatum, a subcortical area, becomes a core operational center in the process of social attraction. Once we see someone who is determined by our cortex to be attractive, the ventral striatum (which includes the dopamine reward system) becomes activated and translates the visual signal into an anticipation of reward, leading to a physical impulse to approach.

The stress, fear, social engagement and social motivational systems constitute the latest thinking on the subject of how our brain-body-mind maintains internal and external regulation.

> Because of these brain systems you are able to feel soothed by your husband's voice or become enraged by his infidelity. They allow us to tell a story in a gesture or a glance. They translate good relationships into a sense of well-being

and robust physical health. They can also take early abuse and neglect and turn them into a lifetime of anxiety, fear and illness.

(Cozolino, 2006, p. 65)

In other words, these four systems collaborate to organize intentionality, motivation and continuity of self-sameness over time. Interpersonal engagement is made possible because of interpenetration of one's subjective state by an empathic other's subjective state.

Self-state regulation

Reflecting on the distinction(s) between neurobiological regulation and self-state regulation does not hearken back to a kind of brain-mind dualism or the minds and material bodies of Descartes (Clarke, 1992). Rather, the unity of brain and mind is reaffirmed. I am pointing out that a process of neurobiological developmental differentiation allows us to draw useful distinctions about certain functions, heretofore taken for granted.

Harter (2012) proposes that there are two perspectives on self-regulation specific to the second decade of life. She first considers the customary variety of adolescent risk taking as an example of the lack of internal regulation, perhaps better put as a state of dysregulation. With the hormonal changes of puberty exacerbating emotional intensity, strong drive expression and sensation seeking, the amygdala becomes more activated, signaling a lack of affective regulation. It is only with the maturation of strengthened cognitive processes, some ten years later, such as the acquisition of abstract reasoning, anticipating consequences, enhanced decision making and problem solving and improved judgment that the developmental lag between emotion and reasoning is closed.

> As Dahl (2004) astutely observes, the intense affects that are associated with puberty can hijack responsible decision making from adolescence into emerging adulthood. This mismatch can compromise the ability to make the kind of reasonable independent choices that society seems to expect, if not demand, of emerging adults.
>
> (Harter, 2012, p. 149)

The issue with risk taking seems to hinge on the degree to which the risk is threatening. "When a stimulus is understood to be nonthreatening, amygdala activation decreases until another potential threat arises" (Cozolino, 2006, p. 167). Young adult regulated functioning becomes possible with the maturation of the prefrontal cortex and its enhanced interconnectivity (Harter, 2012, pp. 148–149). We can now say that very different biological markers, puberty and prefrontal cortical maturation, may be said to act as bookends to the adolescent phase (Bendicsen, 2013; Spear, 2010).

Harter (2012) comments on a second dimension of self-regulation. "Leary (2008) argues that self-conscious emotions evolved more in response to our concern about what *other people* think of us than as a reaction to what we think of ourselves. In particular, these emotions signal fear of interpersonal rejection." From the perspective of regulating our social interactions with others, "self-conscious emotions serve *self-regulatory functions*, in that individuals must monitor and adjust their personal behavior in the service of sustaining positive relationships with the significant others in their social network" Harter believes a superego component, an internalization of the standards of the other, is also involved. Tangney (2003) argues that

> self-conscious emotions serve as a moral barometer, providing salient feedback about our social and moral acceptability and our basic worth as human beings. They guide and regulate our behavior, providing the motivation to adhere to social and moral standards. These motivations do not preclude a concern with violating one's ideals for the self, given that these personal goals are typically defined by the social expectations of others.
> (Harter, 2012, pp. 194–195)

The regulation of the self-state is discussed from two perspectives: 1) that of self-regulation of the adolescent brain, understood as reaching an optimal state of regulation with the attainment of integrative functionality due to the maturation of the prefrontal cortex with its enhanced neural interconnectivity, and 2) that of an intersubjective component, with mutual recognition undergoing continuous differentiation, forming an interpersonal milieu for the constantly recreated self-state.

The biopsychosocial model

When George L. Engel (1977; 1978) proposed the biopsychosocial model, the dominant health framework was the biomedical model. The biomedical approach to disease was characterized by 1) an absence of the consideration of social, psychological or behavioral factors in illness; 2) reductionism, the philosophical view that complex phenomena are ultimately derived from a single primary principle (e.g., molecular, cellular) organized in simple linear cause-and-effect relationships; and 3) mind-body dualism, the doctrine that separates the mental from the somatic. Despite considerable resistance from the medical/psychiatry community, Engel successfully argued for a holistic/general systems theory approach (von Bertalanffy, 1952; 1968). In this model all elements in systems are interconnected across different levels of organization and a change in one affects change in all the others. The simple cause-and-effect explanations of linear causality are replaced by reciprocal causal modes.

> This approach, by treating sets of related events collectively as systems manifesting functions and properties on the specific level of the whole, has

> made possible recognition of isomorphies (or similarities, authors insertion) across different levels of organization, as molecules, cells, organs, the organism, the person, the family, the society, or the biosphere.
>
> (Engel, 1977, p. 196)

It is a very short step from this paradigm to contemporary non-linear dynamic systems theory or complexity theory and its various applications, in particular, those related to regulatory processes. One of the most recent applications in mental health is illustrated by Palombo (2017) and his use of the biopsychosocial framework to explicate his levels of analysis in the neuropsychodynamic treatment of self-deficits. See Chapter Five for a discussion of complexity theory.

Tronick's Mutual Regulation Model

Various models of psychobiological co-regulation are emerging. One model that has many elements consonant with my regulation hypotheses and its operationalization through a developmental algorithm is Tronick's (2007) Mutual Regulation Model (MRM). I will elaborate on the MRM because of its multiple points of significant intersection with my work. The MRM draws on the wide-ranging research of many contributors, especially that of Brazelton and Tronick (1982). "The MRM sees infants as part of a dyadic communication system in which the infant and adult mutually regulate and scaffold their engagement with each other and the world by communicating their intentions and responding to them" (p. 1). The MRM seeks to join dynamic systems theory, with co-created meaning-making and intersubjectivity with the concept that mother–infant interaction has as its goal the achievement of a state of mutual regulation through interactive behaviors that are primarily affective displays (Tronick, 2007, p. 1).

The MRM rests on two sets of propositions. First, humans are makers of private meanings in their purposeful relations to the world. Human functioning is understood as organized by complex systems. Through a hierarchical, multilevel psychobiological array of systems, humans use energy constantly to exchange meaningful information to make sense of their individual place in the world. "This sense-of-oneself in the world equals the totality of meanings, purposes, intentions and biological goals operating in every moment on every component and process at every level of the system from molecules to awareness" (Tronick, 2007, p. 2). "The totality of meanings can be conceptualized as a psychobiological state of consciousness" (p. 2). Set in the context of various levels of awareness the state of consciousness in its broadest terms includes "reflective alertness, preconscious, unconscious, dynamic unconscious, reverie, daydreaming, the multiple states of sleep, meditation, and mindfulness, and even biologically–oriented physiologic, neurologic, endocrine and other somatic states" (p. 3).

> When humans are successful in appropriating meaning into themselves, our biopsychological state moves away from entropy, becomes information-rich,

exists at the edge of chaos, and new properties (meanings) emerge. When humans are unsuccessful, in appropriating meaning, the biopsychological state dissipates, loses complexity, and properties of the system are either lost or fixed.

(p. 3)

A particularly effective way of growing and expanding complexity occurs when two or more individuals convey and apprehend (i.e., take hold of) meanings from each other to create a dyadic state of consciousness.

(p. 3)

This co-created dyadic state results in an expanded capacity for an expanding matrix of meaning, thus increasing complexity.

Second, Tronick's theory of human development includes the following dimensions. The theory must be experience near and clinically authentic. The separate worlds of the researcher and the clinician yield different perspectives requiring the deliberate efforts of keen critical thinkers to distinguish different conceptual orientations or run the risk of conceptual incoherence. Cross-discipline application of concepts requires attention to clarity of definitions and contexts. Conceptual mixing can lead to eclecticism where the outcome can be experience distant, such as in neuroscience models of the brain where the individual is absent (Tronick, 2007, pp. 4–5). Theorizing about human development must locate the individual in the center of the ongoing change process. Overreliance on case histories can lead to shortcut conclusions such as in the "adult end-state" models where, for example, adult borderline conditions are explained through vicissitudes in infant experience. Inferences or conclusions drawn from partial data that isolates or marginalizes the complexity of the totality of multiple factors continuously impinging on development becomes experience distant and compromises clinical authenticity (pp. 5–6). In addition, Tronick recognizes that the effect of culture on development tends to be minimized. Actually, all domains of developmental experience are culture bound through shared symbols and language. The last dimension Tronick emphasizes is that the theory needs to move away from both descriptive and dynamic depictions of development, which can be experience distant, toward the generation of testable hypotheses.

With respect to the MRM model itself, the MRM views infants and caregivers as parts of a larger regulatory system.

The MRM postulates that infants have self-organizing neurobehavioral capacities that operate to organize behavioral states (from sleep to alertness) and biopsychological processes – such as self-regulation of arousal, selective attention, learning and memory, social engagement and communication, neuroception, and acting purposefully in the in the world – that they use for making sense of themselves and their place in the world.

(Tronick, 2007, p. 8)

Following Winnicott (1964), while the infant has an impressive array of capacities, they are only of momentary usefulness without the regulatory input of the mother: "regulation was accomplished by the operation of a communication system in which the infant communicated its regulatory status to the caregiver, who responded to the *meaning* of the communication" (p. 10). The postulate that the infant-caregiver dyad can only exist in a co-regulated, co-created meaning-making exchange spelled for Tronick the end of the theory of the self-contained infant and the end of one-person psychology. Employing Habermas' (1969) concept of intersubjectivity as a communication exchange of meaning, Tronick seems to regard intersubjectivity as a state that contributes the foundation to understanding the dynamic nature of the exchange of meaning and how shared meaning grows the complexity and coherence of the individual: "intersubjectivity is a pre-cognition to meaning-making, states of consciousness, and other dyadic expansion" (p. 11). The MRM considers the infant-caregiver interaction to be developmentally messy and uncoordinated much of the time with the mismatching and reparatory matching of affective states and relational intentions as the primary business of the co-regulating dyad.

Tronick explicates the linkage between dynamic systems theory and co-created narrative in the following account:

> A fundamental principal of the MRM is that the form of the interaction and the meaning of the relational affects and intentions that regulate the exchange emerge from a cocreative process. Cocreative processes produce unique forms of being together, not only in the mother-infant-relationship, but in all relationships. Cocreation emphasizes dynamic and unpredictable changes of relationships that underlie their uniqueness. I want to emphasize that I am not using the term *cocreative* as a reformulated substitute for the more commonly used term *coconstruction*. Coconstruction contains a metaphor of a blueprint that implies a set of steps for getting to an end state. Rather, it implies that when two individuals mutually engage in communicative exchange, how they will be together, their dynamics and direction, are unknown and emerge only from their mutual regulation.
>
> (Tronick, 2007, p. 466)

A narrative account of the exchange before or during the exchange is not possible. Narration can only be created after the experience. Over time as new forms of being together are created, the relationship becomes more unique and differentiated.

Considering the clinical implications of MRM, Tronick suggests that in increasing dyadic states of consciousness there are experiential consequences to the process of expanding the complexity and coherence of such states.

> When new information is selectively incorporated the individual experiences a sense of expansion, joy, and movement into the world. A successful increase in complexity leads to a sense of connection to the other person in the

dyadic state, and a relationship to him or her emerge. Importantly, a sense of connection to one's self develops, accompanied by a feeling of solidity, stability, and a continuity of the self. Additionally, the inherent momentum that comes from forming states leads to an impending certitude about one's place in the world – a sense, in or out of awareness that, 'I know this (whatever it is) to be true'.

(Tronick, 2007, pp. 14–15)

This sense of certitude about "my place in the world" is suggestive of Stolorow and Atwood's (1992) role attributed to self-delineating selfobjects which sustain the self by a process in which the self acquires an experience of the world and the self as real. "'Reality,' as we use the term, refers to something subjective, something felt or sensed, rather than to an external realm of being existing independently of the human subject" (pp. 26–27).

It is our view that the development of the child's sense of the real occurs not primarily as a result of frustration and disappointment, but rather through the validating attunement of the caregiving surround, an attunement provided across a whole spectrum of affectively intense, positive and negative experiences. Reality thus crystalizes at the interface of interacting, affectively attuned subjectivities.

(p. 27)

The self-delineating selfobject function may be pictured along a developmental continuum, from early sensorimotor forms of validation occurring in the preverbal transactions between infant and caregiver, to later processes of validation that take place increasingly through symbolic communication and involve the child's awareness of others as separate centers of subjectivity.

(Stolorow and Atwood, 1992).

It seems to me that two theoretical positions, one from Tronick's co-created meaning-making fueling complexity and the other from Stolorow and Atwood's self-delineating selfobjects contributing to affectively attuned subjectivities, can be usefully linked to further explain the forces at work along the developmental continuum. In the process of the late adolescent becoming a young adult, the mobilization of the Transformational Self exerts an attractor effect in the dynamic self system to create an experiential multiplier that solidifies the meaning of the reconfigured self-state. The reconfigured self-state needs to be consolidated. This is done through acquiring a sense of certitude (Tronick) about my place in the world and a sense of the real (Stolorow and Atwood) about being a distinctive subjectivity in an intersubjective environment. It is as if the newly minted young adult now believes, "I know who I am and I know this to be true." In an ego psychology framework, Blos (1962) referred to this consolidation as emerging with a sense of decisiveness and individuality, giving "tonus to the personality" (see Bendicsen, 2013, pp. 48, 134).

Regulation and resilience

What is resilience? In common parlance it is the idea of a tree bending, but not breaking, in a strong wind typified by Walsh (2003, p. 399) who defines resilience "as the ability to withstand and rebound from disruptive life challenges." Fraser, Kirkby and Smokowski (2004, p. 23) define resilience as "successful adaptation despite adversity." Werner and Smith (2001, pp. 284–285) define resilience as "The self-righting tendencies within the human organism" (in Urdang, 2008, p. 156). Rutter (2006) "has defined resilience as individual variation in the relative resistance to environmental risk experience" (in Tronick, 2007, p. 378).

Resilience is usually discussed in the context of risk factors and protective factors. "Environmental risk factors include limited opportunities for education or employment, racial discrimination and injustice, poverty, child maltreatment, interparental conflict, parental psychopathology, poor parenting." And an insecure attachment.

> Individual psychosocial and biological risk factors include biomedical problems and gender. Environmental protective factors include opportunities for education; employment; growth and achievement; social support; presence of a caring, supportive adult (not necessarily the parent); positive parent-child relationships; effective parenting.

And a secure attachment. "Individual psychosocial and constitutional protective factors include easy temperament, competence in normative roles, self-efficacy, self-esteem, and intelligence" (from Kirby and Frazer, 1997 in Urdang, 2008, pp. 157–158).

It seems intuitive to consider the matter of resilience as a dynamic function integrated with regulatory processes. Tronick (2007) advances the "normal stress resilience hypothesis." It "is framed by a dynamic systems perspective on development of behavior and the brain and the processes that regulate development, in particular the interactive communicative engagements between infants/children and caretakers that regulate stressful experiences" (Tronick, 2007, p. 378).

Let us take a moment to understand what is meant by complex systems.

> Complex systems are systems that have a hierarchical organization operating at multiple size and temporal scales, and they are information-rich with local contextual interactions. Complex systems exhibit emergent properties at different levels that are neither fixed nor chaotic. Self-organizing processes generate these emergent properties and lead to an increase in the complexity of the system, but there are always limits on a system's maximum complexity. Mature and healthy open biological systems are in a dynamic state of organization that approaches these limits, such that the mature organism attempts to garner energy to maintain and optimize its level of coherence and complexity. However, in mature systems emergence of new properties

> is limited and variation in the relative complexity among individuals is primarily related to their success or failure in gaining fitted energy. Further, for mature organisms, self-organizing maintenance of complexity becomes increasing demanding energetically and when there is failure to achieve sufficient amounts of energy, the system begins to dissipate and lose complexity. In the extreme there is a shift to a less complex state or phase.
>
> (Tronick, 2007, p. 379)

This lower state of organization can move toward entropy eventuating in death.

Tronic builds on his empirically grounded Mutual Regulatory Model in his understanding of resilience as an adaptation to normal stress: 1) in a dynamic systems perspective stress inevitably and ubiquitously travels with normal developmental change and its concomitant, interactive regulatory processes govern the stress of that developmental change; 2) as a result of coping with this normal developmental stress the child develops new, more effective coping capacities, enhancing his ability to manage normal and even abnormal levels of stress; and 3) the unique success or failure of each individual's experience of stress accounts for individual variations in coping with normal stressors. This dimension explains differences in individual resilience.

Open biological systems must acquire "fitted" (usable) energy and a mature system must properly utilize that energy to increase its level of complexity and facilitate the ever reconfiguring self-organization. System dissipation is avoided by the formation of a dyadic regulatory system. An extensive series of psychobiological micro exchanges allow developmental synergies to emerge to further drive complexity. A major consequence of the dyadic regulatory system is that over time the capacity for self-regulation expands, requiring less external scaffolding. Research suggests that there may be a gender difference with boys whose mothers are depressed, having a more difficult time developing resilience than girls of depressed mothers (Tronick, 2007, p. 392).

Normal interactions are inherently stressful and trigger the sympathetic autonomic nervous system to manage arousal levels. One such stressor is illustrated in the still face experiment in which the mother 1) has an episode of normal face-to-face interaction followed by 2) a still face episode (non-reactive gaze aversion) and ending with 3) a reunion episode and reparation. Neurophysiological reactions are observed in the infant such as increased heart rate and respiration, perhaps set in motion by cortisol reactivity. The parasympathetic autonomic nervous system mediates this stress and returns the organism to pre-stress conditions.

> Critically, the process of mismatch and stress, reparation and the reduction of stress literally occurs thousands of times in the course of a day and millions of times over the course of the year such that as the microeffects of matches, mismatches, and reparation accumulate, they have profound effects.
>
> (Tronick, 2007, p. 389)

The development of the capacity for resilience originates in the repeated sequencing of matching, mismatching and positive reparation. Experiencing stress in the context of healthy social interactions, where stress is not overwhelming, allows for the consolidation of 1) a robust positive affective core, 2) a sense of trusting the caretaker as a reliable partner, 3) a representation of the child as an effective agent in adapting to stress and 4) establishing the narrative that "I can deal with adversity" (Tronick, 2007, pp. 392–394). In the case where stress is overwhelming and traumatic, resilience is compromised. This can leave the child with inadequate, brittle defenses, a rigid stance toward adaptability and, perhaps most damaging, abandoning the vulnerable child who struggles alone to find people to use as selfobjects to help soothe, calm and regulate stress.

Jay (2017) cites the work of Fredrickson (2001) in searching for the central dynamic in resilience. Exposure to positive experiences in the context of love are reparative and combine to create an "undoing effect."

> Love defies simple definition, to be sure which is why theorists like Fredrickson tend to use the word almost as an umbrella term for the good feelings we experience – joy, gratitude, contentment, interest, hope, pride, amusement, inspiration, awe – in the context of a mutually caring relationship. Love, then, carries the double advantage of positivity plus connection, offsetting the harm of both stress and isolation. Acting something like a medicine, it downshifts our bodies and minds and it speeds our recovery from hard times. It may sound like the stuff of clichés to say that love heals, but love has perhaps unequaled power to mend the strain and trauma that have come before.
>
> (Jay, 2017, p. 295)

In the case example cited in Chapter Three, we will underscore the healing power of loving, "mutually caring relationships" in Myles' treatment journey back to healthy functioning.

Schore's neuroendocrinology regulatory hypothesis of boys at risk

Before moving on, I want to highlight another effort at understanding the regulatory processes of early development. This one comes from the exciting work of Schore (2017) and his explanation of why boys are at greater risk than girls. Exploring the deeper psychoneurobiological mechanisms of fetal life, Schore creates a model that adds to our understanding of regulation theory.

> The central thesis of this work dictates that significant gender differences are seen between male and female social and emotional functions in the earliest stages of development, and that these result from not only differences in sex

hormones and social experiences, but in rates of male and female brain maturation, specifically in the developing right brain.

(p. 15)

the stress regulating circuits of the male brain mature more slowly than those of the female in the prenatal, perinatal, and post-natal critical periods, and that this differential structural maturation is reflected in normal gender differences in right brain attachment functions.

(p. 15)

In addition, this maturational delay contributes to greater vulnerability in males in the social environment (attachment trauma), in the physical environment (toxins contributing to endocrine disruptors) and in gender-related psychopathology. Males have higher rates of autism, early onset schizophrenia, attention deficit hyperactivity disorder and conduct disorders (p. 15). This pattern has obvious implications for the later stages and states of development for those of us interested in diagnosing and treating emerging adults.

In concluding this section, the reader is invited to review Bruce Perry's "SevenSlideSeries: The Human Brain" for a concise overview of the brain's remarkable regulatory properties. YouTube access makes this educational opportunity readily available. The SevenSlideSeries began in 2001 and is sponsored by *The Child Trauma Academy*, a non-for-profit organization based in Houston, Texas.

Let us now turn to a case example of adolescent onset schizophrenia. This example will illustration the usefulness of regulation theory as an organizing principle.

Note

1 The insula cortex with its functional emphasis on the congruence of inside and outside experience assumes a vital role in regulating interpersonal relations.

> In a report of a study that was published recently in a top-tier psychiatric journal, researchers described a stunning finding that challenges the notion that there is a plethora of psychiatric brain disorders. They conducted a large meta-analysis of 193 published brain imaging studies of people with schizophrenia, bipolar disorder, major depression, obsessive compulsive disorder (OCD), anxiety and addiction. They found that those supposedly 6 discreet illnesses are all associated with a varying degree of shrinkage (atrophy or hypoplasia) of the same three brain regions.

(Nasrallah, July, 2015, p. 10)

They are the dorsal anterior cingulate cortex, the left insula cortex and the right insula cortex.

The dorsal anterior cingulate cortex, located "around the frontal region of the corpus callosum, controls rational cognitive processes, reward anticipation, decision making, empathy impulse control, and emotional response."

The insulae are the cortical regions deep inside the lateral sulcus, which is the fissure that separates the temporal lobe from the parietal and frontal lobes. The functions of the insulae include consciousness, emotions, perceptions, motor control, self-awareness, cognitive functioning, and interpersonal experience.

(p. 10)

These three regions together manage high-level executive functions such as working memory, reasoning (planning and anticipating consequences) and flexible thinking. In addition to the finding of similar anatomical anomalies in three brain regions in the abovementioned six mental disorders, genome research is suggesting a similar genetic substrate in five psychiatric disorders: schizophrenia, autism, bipolar disorder, major depression and attention/deficit – hyperactivity disorder. The data suggesting genetic and structural brain similarities might explain 1) frequent comorbidity of certain DSM psychiatric disorders, such as depression and addiction in schizophrenia; anxiety and OCD in bipolar disorder; depression with OCD in addictions: and so on; 2) "the presence of intermediate phenotypes in unaffected family members, such as cognitive dysfunction in the parents of patients with schizophrenia, compared with parents of matched healthy controls"; 3) "the much higher rate of psychopathology among family members of patients with a major psychiatric disorder, compared with the general population" (p. 11).

While psychiatrists may agree that the symptoms of the six disorders mentioned in the beginning are very different, all agree that the core cognitive function that has been compromised is flexible thinking. Inflexible thinking might manifest as a core feature in schizophrenia with its paranoid or implausible delusions, in bipolar disorder with grandiose delusions, in major depression with a fixed false belief of worthlessness and/or hopelessness, in anxiety with the fixed false belief of impending doom or death, and in OCD with ego-dystonic false beliefs (such as obsessions) that can progress into ego-syntonic delusions (p. 11).

3
CASE EXAMPLE

As the case formulation literature expands to include neuroscience hypotheses and a new category of speculations about the human experience, we are witnessing a spate of books on the subject with a trend toward the topic of neurobiological regulation (e.g., Siegel, 1999; Schore, 2002; Cozolino, 2006; 2010; Montgomery, 2013). This chapter contributes to this stream of neuroscience formulations by citing a case in my own practice of over ten years duration. It is an example that uses the metaphor of human experience in conjunction with neuroscience functionality to explain a serious dysregulation condition. My practice orientation can be considered both psychoanalytically informed and neurobiologically oriented. My thinking about case dynamics is organized by a process captured in the flow chart entitled "Critical thinking mental health decision-making flow chart" which can be found in Appendix I. I believe the following narrative captures the richness of human experience without an overbearing reliance on the jargon of neuroscience. I want to thank Myles for allowing his moving story to be told. Identifying details have been altered to protect his anonymity.

It should be mentioned that in the *Transformational Self* (2013) I demonstrated regulation theory through two case examples of adolescent girls. In this exposition I enlarge the applicable population to include adolescent boys and those with serious mental health disorders. So the Transformational Self hypothesis is gender neutral and relevant to a wide range of conflict and deficit conditions on a developmental continuum.

The following case example is intended: 1) to illustrate the dynamic formulation possibilities found in a developmental algorithm, 2) to broaden the application of the Transformational Self's clinical usefulness by examining the interrelationship between the self-referencing metaphor and the self-regulating metaphor, 3) to give the reader an appreciation of the continuing vicissitudes of treatment in this complex case, 4) to explicate the role that "mutually caring relationships" play in the development of resilience and 5) to recognize our patient's strengths to share their story and, in so doing, give all of us so much insight.

This treatment experience takes place in two phases. The first phase lasted two and a half years and began as Myles was halfway through high school and ended abruptly six months after graduation. The second phase began eleven months later and, to date, has lasted over eight years. It continues through to this writing. Myles has just celebrated his twenty-fifth birthday.

Myles' journey – phase one

Myles was referred to me in his junior year by his high school counselor who believed Myles was having a series of delayed negative reactions to his parents' divorce. Myles was 13 and in the beginning of seventh grade when his father suddenly left. Myles was the younger of two siblings; his sister was six years older, a high school senior. Myles' developmental history can be said to fall within the average expectable environment range. His sister was developmentally on track and achievement oriented. Both parents were employed in different fields and had created an upper-middle class family set of circumstances when the divorce occurred. While each parent was free of mental illness, on each side of the extended family there were relatives struggling with Axis I mental disorders. Of the two children, Myles took the news of the separation and divorce the hardest. Myles remembers his father gathering the children on a holiday weekend saying he had an important announcement to make. The children, sensing something ominous, sat on the stairs connecting the second floor bedrooms to the ground floor of the house. His father and mother were in the adjacent living room. His father said he was tired of being married and was getting more and more unhappy. He was going to move out to live by himself to sort things out. His mother seemed resigned to the news and said sadly, "I can't change your father's mind." Myles had no awareness of the difficulties between his parents. He was shocked and stunned into silence in contrast to his sister who peppered their father with questions. His father soon moved out without discussing matters further with Myles.

Two years later, Myles entered high school and quickly found himself adrift. Toward the end of his freshman year, one of his teachers noticed a certain sadness in Myles' demeanor, a lack of interest in getting good grades and a change in peer relationships. He was losing interest in school, began sampling marijuana, participated in basketball, but soon dropped out. He continued, however, with his long-time interest in drums, forming a band and associating with a musical drug culture. By now his sister was off to college, leaving Myles and his mother continuing to live in the family house and adapt to the finality of the divorce. It seemed to Myles that his father became absorbed in his new life and interests and avoided his former family.

Late into his sophomore year at age 15 years 10 months, Myles and I began our work on a once a week basis and the parents with Myles were seen monthly. The parents remained cordial and dealt with each other in an amicable manner, both placing the welfare of Myles first. Myles' depression was obvious. He formed a solid therapeutic alliance and used the relationship in a substantial way to come to terms with the divorce. In family therapy Myles found his voice and reached out to his father, improving their relationship. With his sister off to college and work, she never participated in family therapy. His father upheld his financial obligations to his family and has become a dependable figure to Myles. Myles obtained his driver's license and was a careful driver. In addition, the band expanded in importance in Myles' life. Even though Myles was the youngest member of the band he was one of its strongest leaders in helping to arrange gigs, compose songs and play drums; Myles said proudly, "I keep the rhythm." This was the first self-referencing metaphor he used. It referred not only to maintaining musical rhythm, but also to being the glue that held the band together. When Myles wanted to take the family car to haul band members and equipment to a gig some 100 miles away, both parents balked. Myles had secured his driver's license only a few months earlier. At the gig, it was anticipated by all that there would be drug use in the audience; Myles nevertheless defiantly declined his father's offer to drive the group, worried about the perception of needing to have a babysitter. After extensive discussion a compromise was reached. His father would drive the group, but not sit in the audience during the performance. His father would neither supervise the group nor observe possible drug use in the audience. This experience was successful as judged by all parties. The ability of his father and Myles to reach this accommodation was seen as a strength, as well as a credit to his father's recognition of the need to support his son's autonomy strivings.

Six months after graduation from high school Myles abruptly stopped coming. The basis for this decision was not made known until the beginning of the second phase.

Myles' journey – phase two
Month 1 to month 17

After an absence of about eleven months, Myles, now 19 years, 2 months old, called and asked if he could resume treatment. I soon learned that Myles withdrew suddenly from treatment because his father decided that Myles had experienced enough help and now should be able to "stand on his own two feet." Myles believed his father meant for him to stop immediately; his father's version was that Myles should begin to discuss in therapy what further gains could be expected and might it be time to begin the termination process. Myles had been taking a course in a local junior college and struggling to learn, found it difficult to read and absorb the material. He later took an online course and found that particularly difficult. He had experienced a frustrating time searching for a job and finally landed one as a sales clerk in a large home improvement chain store. He was struggling with performance expectations and was worried he would be terminated. He was depressed and having severe difficulty functioning. Also, Myles disclosed that he had been having very troublesome visual problems that he had concealed from me and his family, saying he did not want to burden anyone. The visual disturbances began in sixth grade and gradually worsened. By his freshman year the visual symptoms had progressed to the essential form that would last. The visual symptoms consisted of a fixed set of horizontal wavy lines, in an amplitude configuration, parallel to each other. The pattern never abated, but it was broken when looking at a face or a moving object such as automobiles in traffic. Without apparent trigger the pattern of visual snow, as it has become known, could intensify to an opaque blur of static or snow. This fearful condition was terrifying and, when combined with Myles' severe depression, left him depleted and exhausted.

What followed were a long series of consultations and tests including appointments with various medical doctors, eye examinations, an EEG to rule out seizure activity, a brain scan, a sleep study and a neuro-psychologist and finally a neuro-ophthalmologist. In the sleep test it was determined that a tonsillectomy was needed to improve what was a restless sleep pattern. That procedure was successful. Anti-depression medication was prescribed, but the depression deepened and the visual problem remained undiagnosed. Myles had been scheduling most of these tests himself and had grown frustrated and disappointed at the inconclusiveness of this laborious process.

The disorder progressed and fulminated (in the sense of exploded) in a panic of terrifying, out of control behavior and thoughts. After about eight months of such unstable, oscillating activity, Myles experienced a devastating collapse with intense suicidal thoughts accompanied by a mixture of severe depression, anxiety and hallucinatory symptoms. He thought he was being followed and was hearing people call out his name behind his back. He was psychiatrically hospitalized for three days where he was diagnosed with schizophrenia and suicidal depression. After three days Myles found this intervention so unpleasant that he insisted on being discharged. He vowed to never enter another psychiatric unit.

As a result of this episode this therapist referred Myles to a biological psychiatrist who confirmed the diagnosis and added "with possible bipolar features." This later was modified to schizoaffective disorder with bipolar involvement. Trial exposure to a wide variety of psychotropic medications followed. Different chemical combinations were tried in an attempt to stabilize his condition. Myles was severely dysregulated. With concentrating so difficult, he struggled to continue driving, generating anxiety for all concerned. Proud of his good driving record, he, nevertheless, hit the curb three times, requiring replacement wheels, but managed to avoid accidents. With intense episodes of depression, anxiety and auditory hallucinatory activity, powerful suicidal thoughts emerged, subsided and reemerged. While uncrating inventory, Myles became fearful he would cut himself with a box cutter, and so attempts to return to work were unsuccessful. A four-month paid, short-term disability leave was secured. He grew to feel so unsafe that living at his mother's home at times left him too vulnerable. Fearful he would overdose or cut himself, he went to his father's home for protection from himself. His father reassured Myles and slept in the same room, allowing Myles to finally sleep.

Concern for his safety and his fluctuating condition prompted two further attempts at psychiatric rehospitalization; both were found to be unnecessary by the admitting doctors. It was felt that the support system in place could safeguard Myles, given his present circumstances. A five-part wellness plan with specific responsibilities for each participant was implemented and reviewed periodically. It included a safety plan, a job/school component, a relationship/recreation section and a self-care part including exercise, diet, self-grooming and developing a healthy sleep pattern. The wellness plan was considered essential to facilitating a return to self-regulation. Family sessions were vital and Myles continued in twice weekly psychotherapy. His father accompanied Myles for the medication monitoring consultations to assure that the psychiatrist had the fullest account of the effects of the medication. Both parents communicated and cooperated well during this difficult period.

At about the fifteen month mark, two events occurred. Myles' eye condition was finally determined to be an unusual feature of schizophrenia. Myles' report that area rugs and hanging pictures could rotate, combined with the optic wavy lines, suggested a rare visual, neurological hallucinatory process that eventually was labeled "chronic, atypical visual distortion." Also, the current medication mix finally began to stabilize his condition. While still very tired, he resumed a more normal sleep pattern. Auditory hallucinations and the rotating objects have abated. Myles has been able to work actively and collaboratively with the psychiatrist to monitor the effects of the medications and adjust accordingly. Presently, Myles is taking Lithium for bipolar, Zoloft for depression, Fanapt for psychosis, Busperidon for anxiety, Klonopin for sleeping and Cogentin for side-effects such as body tremors. His mother supervises the administration of the medication and assures that supplies are adequate. His father stays involved with Myles, accompanying him to psychiatric consultations and playing tennis and golf, with occasional outings at major sporting events. On a scale of 1 to 10, with 10 being the worst, depression has remained at the 3–4 level with no suicidal thoughts; visual distortion has remained steady at level 5, with no hyper-static intrusions; auditory hallucinations have retreated and have remained at level 0–1.

This condition has remained increasingly stable for the past four months. How has Myles found the resilience to persevere and improve his quality of life? Recognizing he cannot accommodate the classroom situation and the academic demands for studying and reading, he has invested in his job with some notable success. He knows he will need to develop compensatory mechanisms to balance out or offset the triple handicap of the full impact of the whole disorder complex in general, his visual syndrome, in particular, and the side-effects of his psychotropic medications. He has made an adaptation to the reality of his situation. He works at being engaging and successful with customers and takes satisfaction in surpassing the productivity of his fellow sales associates. In this highly competitive environment, Myles frequently meets or exceeds his weekly sales goals or metrics. In month sixteen, with considerable relish, the second self-referencing metaphor materialized; he labeled himself the "Top dog on the floor." In month seventeen he was in the lead for a promotion to an entry level management position involving circulating through the other departments giving suggestions and otherwise encouraging floor personnel to enhance sales. Due to slumping sales, management has not filled this position, but Myles remains optimistic about his prospects. In his annual review he received a rating of A-. He has maintained a regulatory type of relationship with his girlfriend, Cindy, who is pursuing a university degree and is a reliable, calming selfobject. Cindy also monitors Myles' cigarette and alcoholic consumption and helps to keep him at the level of half a pack a day and a few beers on the weekend. On one occasion Myles asked Cindy to accompany him to the session. Myles was in a stupor due to the medication and was drooling. Cindy wiped away the drool and looking at me she said, "You know this is not a normal romantic relationship."

Phase two of the treatment has continued, twice a week, into the tenth year. The vicissitudes of the treatment process have been highly varied, intense and unpredictable. Through it all Myles, in a burst of expansiveness, now considers himself "Top dog in the store."

Discussion

Issues in the two phases

In the first phase of treatment, the psychotherapy, with the exception of the abrupt ending due to his father's wish for his son to move on, so to speak, and concealment of the visual distortion, could essentially be considered successful along traditional lines of goal attainment. Myles became fascinated with his dream life and used his dreams to understand better his internal life with considerable satisfaction. Myles reconciled to his parents' divorce, worked through the loss of a romantic relationship, developed an internal motive for learning, restricted his peer relationships to healthy friendships, invested in a band with some success and graduated with a determination to engage the next phase in his life. This therapist was left with a feeling of puzzlement and dissatisfaction over the abrupt ending. Should this termination "process" be considered an interrupted treatment due to an expression of an autonomy striving, fear of withdrawal of support from his father, an unconscious repetition of his father's withdrawal from the family during the divorce and/or an acting out of ambivalence toward the treatment?

The traditional search for the explanation of the rupture in treatment involves a microscopic search for issues and/or flaws in the patient-therapist alliance such as an empathic failure. However, in the past decade, another approach has emerged to clarify possible causes associated with the breech. Dynamic systems theory is being used increasingly to explicate development- and treatment related issues. Galatzer-Levy has written a series of papers on the subject of non-linear psychoanalysis. (See Chapter Five for a brief discussion of complexity theory.) Using a dynamic systems approach an alternative hypothesis for the interruption presents itself. In a non-linear system changes can be abrupt due to unseen attractors, or organizers. These changes should be understood as system features devoid of human motivation (Galatzer-Levy, 2017, p. 78). The overall process of change should be examined and conceptualized as being in a state of disorganization or rather reorganization, on the edge of chaos, as a new, reconfigured state emerges. The overall system complexity becomes richer, allowing for creative solutions to problems. Considering both phases in treatment as one ongoing process enables us to see the creation of other attractors in a coupled oscillation of different possibilities, namely the emergence of phase two (pp. 87, 105, 163).

In the second phase of treatment, the emphasis shifted to self-regulation. External supports were put in place with a suicide prevention/intervention plan and supportive treatment elements with clear responsibilities for parents, Myles and therapist. During the first fifteen months of this phase, circumstances were exceptionally fluid and turbulent. As the disease process erupted, a profound sadness overtook this therapist due to feelings of uselessness and inadequacy. The intense exposure to such primitive material had taken its toll. The suffering Myles was experiencing could not be articulated and I felt unable to help. Fortunately this therapist had been a member of a long-time study group which provided vital consultation that helped to regain a therapeutic perspective. Individual consultation with a colleague supplemented this effort. Cases of such difficulty require careful consultation to keep the therapist focused and grounded. As we grew to know the nature of his disorder and it stabilized, it seemed to us that it could be managed and sprouts of hopefulness reemerged.

Issues receiving the most attention were 1) the need to stay self-regulated in keeping appointments, getting to work on time, self-grooming, and so on and 2) managing anxiety associated with work. As continuing in school became impossible, his core identity, self-esteem and self-confidence began to be more and more closely linked to his job success. Reliable persons were found to help him stay self-regulated. These persons served as selfobject experiences in that they helped modulate intense, fluctuating affect states, cushion frustration, bolster a battered self-esteem and demonstrate that he was loved and cherished. Myles' regressive dependency fears were offset by validating autonomy strivings that were supported by his father, as was the case when his father accommodated him in the earlier band driving incident. Improved coping strategies and skills were tested, but remained in place as the disorder fluctuated.

Bollas (2015) has described in great vividness the point at which the frank schizophrenic process manifests:

> When the breakdown occurs, the self loses the function of historicity – the capacity to transform the past into a narrative. The loss of historic capability happens partly because the schizophrenic's mind is no longer capable of this kind of integrative work, and partly because contacting the past is too painful.
> (p. 88)

Three aspects of the breakdown process are noted:

> These activities – the eradication of one's history, the invention of a personal mythology, and communion with the thingness of the world – can be floridly

disturbing to the other, but at times there is an indescribable sweetness to the occasional schizophrenic inventiveness.

(p. 89)

"Thingness" refers to obsessions about objects such as rocks, trees, bicycles, and so on. While Myles was teetering on the rim of the schizophrenic abyss, he never, at least as judged by the coherence of his speech, lost psychic integration. He was able to always locate himself in the intersubjective space of experience. Hallucinations, both visual and auditory, were always understood as bewildering, alien and these terrifying, episodic mental contents never dominated his existence. The keys to recovery are to be engaged long term with others and to be able to talk about the psychotic process. Fortunately Myles was able to make good use of psychotherapy and continued to co-construct a self narrative of resilience and survival.

Resistance to twice weekly treatment grew and we negotiated reducing the frequency to once per week. Myles was saying he was feeling better and could manage his life better than before. With the shift to weekly sessions, Myles' participation in treatment has improved. The movement toward a promotion was understood as remarkable progress, a testament to his initiative, his relationship skills and confirmation of his resilience. Might this be a repetition associated with the autonomy strivings Myles successfully tested during the driving of the band episode?

Month 17 to month 49

Seldom has a case, in my experience, demonstrated so many unexpected twists and turns. Developments of such a magnitude have occurred that a continuation of the case record is warranted. I want to pick up the narrative of Myles' story in the seventeenth month of phase two treatment to month forty-nine.

Cindy had been responding less and less to Myles' communications and her messages were getting briefer. Clearly something was shifting in Cindy and it was evident that she was decathecting the relationship. Myles began getting messages that perhaps it was best to move on to other relationships and experiences, while still remaining friends. This was very difficult to accept, but Myles' slowly adjusted to the reality of Cindy being a person who exchanges an occasional perfunctory text message and little more. The relationship had lasted about three and a half years, bridging the two phases of treatment. The question now was who or what was to replace her vital, lost selfobject functions of soothing and calming?

In a second development, as the relationship with Cindy was ending, the relationship with his father was experiencing change. Myles' father remarried and relocated some twenty-five miles away. Myles' second online course ended with a failing grade. Myles was reluctant to tell me the grade and even more hesitant to tell his father. Feeling so ashamed as an academic failure, how could he explain to his father, that the $500 his father spent on tuition was wasted? How could he approach his father for more tuition money? How could he explain that he now had a massive learning disability as a result of the atypical visual distortion?

A third development also emerged. In the thirtieth month Myles abruptly announced that he would be interrupting treatment to check into an alcohol rehabilitation facility in Florida! This was an absolutely astonishing announcement. Myles left two days later. I heard nothing until two weeks later when Myles called to say he was back and wanted to resume treatment. Myles had discussed with his father the full scope of his drinking and that both agreed an intervention of this magnitude was required. I was completely out of the loop and blindsided by both the disclosure of alcoholism and the decision to seek rehabilitation. This brought to mind the earlier, sudden interruption of treatment ending the first phase and the long-held secret of the visual distortions. What was driving these periodic "pockets of privacy"?

I met with Myles and his father before resuming individual treatment with Myles and the following story emerged. Myles was being processed into the residential treatment program and was undergoing a wash of psychotropic medications when he "blacked out" or "spaced out." He understood later that his loss of consciousness was the result of an overdose of "benzos" used to stabilize alcoholics in the wash process. He awoke in a hospital emergency room and was told that when he was admitted he had no pulse, actually had died, and was shocked back to life. After a day in the hospital he was returned to the rehabilitation program and was shortly thereafter discharged as requiring a different type of program. He was told that records would be sent, and after multiple requests, none were ever received.

Back home, his biological psychiatrist immediately admitted him to a six-week outpatient hospital-based alcoholism program. Myles also began participating frequently in local Alcohol Anonymous meetings and became attached to both processes. Myles cooperated fully, completely stopped drinking, and was placed back on his customary mix of six psychotropic drugs. About five months before Myles' decision to go to Florida, he had contracted pneumonia and had secured a medical leave from his employer to recuperate. By the time he left for Florida his "lungs had healed," but he remained on medical leave until he returned to work three weeks later. Myles' condition rapidly stabilized with the plan that he was to return to work within thirty days, if not sooner. The treatment plan was revised with the additions of placement on a suicide watch, psychotherapy increased to twice per week, successful completion of the outpatient program and attendance at AA meetings with a frequency of three to four times per week. The family support system was reinvigorated, Myles returned to work and monthly family treatment sessions were resumed. It was stressed that Myles develop a reliance on his AA sponsor, resulting in telephone contact at least once per day coupled with occasional interaction at meetings.

Our sessions were filled with the themes of working the plan, his near death experience, the history of his drinking, returning to work in the knowledge that fellow employees are aware fully of his alcoholism and partially aware of his overall condition, the omnipresent fear of relapsing and the motivation for and the impact of the disclosures of these "pockets of privacy." He contrasted the differences between selling drunk versus selling sober, as he described his current sales technique. "It's nice to be counted on and people are not avoiding me," he said. At about the ninety day mark of sobriety, he experienced going through a phase of struggling with his cravings for alcohol and considering securing a prescription for vivitrol, a widely used anti-craving medicine. After about a week, this urgency passed and he proudly celebrated his anniversary of one hundred days of sobriety. This milestone took on immense significance and seemed to embolden Myles to aspire to become a drug and alcohol counselor. He reached out to a leader of his old church youth group, who also works with high school students, and she encouraged his networking efforts. Proud of his steady progress, he wanted to speak to at-risk youth about his experiences and is highly motivated to pursue this goal. Recently, he said he was more aware of his hygiene, especially when getting ready to go to work.

In the thirty-sixth month mark in treatment Myles approached the seven-month mark in sobriety and reported he joined a "Big Book" study group. Each member takes turns reading passages. In an achievement of adaptation, Myles said reading was becoming easier, suggesting an accommodation was being reached where compensatory reading processes are evolving.

In the thirty-ninth month mark in treatment Myles announced that his supervisor was planning on promoting him to the store-wide inventory control manager. After an exhausting holiday season on the floor, Myles had been performing extra tasks not directly related to retail sales. Myles said, "Somebody had to do it; to not do it made the department very inefficient." His "above and beyond" efforts were noticed and the expectation was that he would take this attitude store-wide. While these expansive efforts were noted, Myles was cautioned not to neglect his core responsibilities. In the fortieth month in treatment, at the age of twenty-two-and-a-half, Myles was now anticipating adjusting to a management perspective with all that that entailed for his new identification as a member of the management team. The "Top dog on the floor" would now become the "Top dog in the store."

However, over the next six months (through the forty-sixth month) the promised promotion did not materialize. His supervisor said that anticipated mid-level staff turnover had slowed delaying the redistribution of staff. In addition, sales were in a slump, placing all personnel transactions on hold. Myles was experiencing progressively more and more stress and frustration during this period. Staff shortages required him to work a string of sixty hour weeks until new hires were able to work the floor. Slippage was noted in his management of medication. He began to be more vocal with his supervisor. With so much at stake the need for discretion and good judgment were paramount despite his state of exhaustion. All communication vibrations from management pointed in the direction of the promotion. In month 45 of treatment Myles was given one of four keys to the central stock room, sending a message to all team members about the confidence leadership had in him. His supervisor told him his situation was the subject of ongoing upper-level management discussions. While he had demonstrated outstanding performance in a wide range of sales metrics, he had no significant experience in managing customer complaints and problem solving. A special position was being considered in merchandising and pricing, respecting his request to get him off the floor with its tiring and repetitious direct sales responsibilities. Myles felt his contribution was being appreciated and recognized and his patience increased. During this time period Myles passed his five-year employment anniversary and his one year of AA sobriety. The feeling he had of success was palpable. In the forty-fifth month mark in phase two treatment, remarkable steadiness in the face of adversity had been demonstrated and Myles was finally feeling his goal was within his grasp.

In the forty-sixth month the move to a new position was finalized. While it was not the inventory control position first envisioned, it was the merchandizing and pricing position. Myles was pleased to receive the news, but his joy was restrained because the lead title for that position was not forthcoming. He received a change to a fixed schedule and an hourly increase, but the title was withheld, he felt due to his inexperience. To signify the significance of and to celebrate the change, I gave Myles an Avo Nuttorno cigar which he thoroughly enjoyed with his friends. The next month revealed the stress that accompanied the new position. The stress was taking a toll on him and his frustration was evident. "I guess I have more on my plate than I bargained for," he said. He struggled to keep on top of merchandise signage changes and periodic reports and grew frustrated and overwhelmed. The "Top dog on the floor" had met its match, he said. The potential for regression was keen. His grandiosity was undergoing attenuation.

Three areas were of especial concern during this period: his personal hygiene, his customary collegial professionalism and peer social interaction. It is well known that an individual with this condition, especially when under stress, is vulnerable for impaired self-assessment. Awareness of personal grooming, referred to as his "presentation," often is compromised (Harvey and Pinkham, 2015). The risk and consequences of being socially offensive were discussed in detail in candid terms. It is hard to rise up the business hierarchy when one smells like a barnyard goat. Myles, when questioned, said, "I need to hear this." The second issue concerned the relaxation of his professional communication demeanor. His frustration gave way to a blunt e-mail to his supervisor about his perception that his supervisor was not following through on his (the supervisor's) share of assignments. The risks of this e-mail being shared up the chain of command could be damaging to his career. Fortunately, the e-mail was dismissed by his superior as insignificant and Myles immediately appreciated the bullet he dodged. The third issue, good peer social interaction, is essential because Myles needs the occasional help from peers to complete his tasks. Should he become abrupt or insensitive peer collaboration would be compromised and his success could be imperiled.

In the forty-ninth month a girlfriend emerged, Sabrina. They met by accident three months ago having been introduced by a mutual friend at an extended family function. After mutual self disclosure about past issues and current struggles, they have become an exclusive couple as she finished college with a degree in elementary school physical education. Her existence was kept a secret in our work and contained in yet another pocket of privacy. After two months of dating he disclosed the relationship, saying he wanted to be sure about her before she was brought into our work. Of the many facets to this relationship, the one that has received the most attention is the nature of the selfobject function she performs in the relationship experience. On the first overnight visit to her college dormitory he neglected to bring a grip with a change of clothes and sundry toiletries, in particular a toothbrush. The next morning, she told him to gargle and brush his teeth. He told her he did not bring a toothbrush. She told him to put some toothpaste on his finger. It soon became apparent that she serves a vital regulatory function. He reminds her to do her homework and do it well. So the relationship has become one of mutual co-regulation. She is in touch with his three vulnerabilities: the need for daily self-care (including medication management and abstinence from alcohol), thoughtful communication and appropriate peer social/collegial interaction. It appears that Myles has found an outside of family and work selfobject experience that serves to secure and strengthen the gains he has made, one that should also equip him for sustained romantic success.

Discussion

The drinking pattern occurred over five years beginning at age seventeen. It continued through graduation from high school and steadily accelerated. He started with sampling rum and coke which gradually led to a preoccupation exclusively for beer. The dynamic driving the drinking during its intense phase seemed to be the threat of losing vital selfobject experiences. The threat intensified after the actual loss of Cindy, a once essential, predictable partner, and accelerated with the impoverishment of another vital self-regulating force, the reduction in contact with his father. The alcohol, then, was understood as a replacement selfobject experience, one that could restore a haze of stability and numb an existence too painful to face.

Myles' rehabilitation involves embedding him in a network of reliable self-regulating individuals, including his AA sponsor, the AA meeting process and the validation of his self-worth. And he continues to be vigorously embraced by family members.[1] These individuals are experienced as restorative selfobjects in that they renew a sense of vitality, hopefulness and acceptance. Myles' alertness and overall availability to use the individual twice-a-week therapeutic process is optimal. The supportive, consistent psychoeducational approach fits well. The expected and customary cognitive distortions and dislocations associated with his schizoaffective condition have not appeared. On a ten point scale, he rates his depression as 2, his visual distortions as 5 and his auditory hallucinations as 0. Personal hygiene, self-motivation and regular attendance at AA meetings are areas of functioning receiving constant attention.

In his twenty-third year his remarkable resilience remains an inspiration and a model for all those struggling for self-regulation. The "Top dog on the floor" self-referencing metaphor now serves as a self-regulating metaphor strengthening his resilience and channeling his capacities toward his goal. His frustration over being a "simple" retail floor associate is being replaced with the fear and anxiety of the possibility of failure in his new position. In a world filled with fresh work challenges in a clear risk-and-reward adult environment, Myles has recognized and realized opportunities for enhanced identifications, as well as the awareness of pathways toward potential developmental derailments.

One of my readers, J. Colby Martin, reminds me that Wolf (1988, p. 27) stressed the importance of sufficient selfobject responses to ward off feelings of coming apart or fragmenting in an experience-near context. For we who work with the most vulnerable among us it is well to remember the absolute need for self-sustaining selfobject experiences to maintain the resilient cohesion of the self.

Month 49 to month 56

This time period covers the therapeutic vicissitudes associated with months forty-nine through fifty-six, approximately four to four-and-a-half years into phase two treatment. In month fifty-three Myles unilaterally stopped taking his antipsychotic medication, Risperdal, saying he did not like the weight gain and the way it made him feel: "like kind of zoned out." Invega was the most efficacious antipsychotic medication he was taking prior to Risperdal, but the cost to obtain a consistent supply became prohibitive. There has been no return of schizophrenic symptoms, such as auditory hallucinations. Our discussions have centered on the clearly defined presentation of bipolar cycling which occurs three to four times per day. Current medications include Cymbalta, lithium, Cogentin and melatonin. He admitted that he forgets to take his lithium in the morning perhaps half of the time! With melatonin he falls asleep easily, but awakens after perhaps two to three hours and cannot stay asleep, waking up perhaps every hour until it is time to rise. He is left quite drowsy during the day. Perhaps the manic-depressive cycling affects his sleep pattern. His girlfriend, Sabrina, has proven to be a dependable selfobject regulating presence. Myles has taken on store-wide training responsibilities for computer processing training, in addition to his regular merchandising and pricing department duties. He is anticipating a formal promotion. He is feeling increasingly stressed and fears dysregulation. He speaks to his AA sponsor perhaps a few times per week and no longer attends AA meetings.

During month fifty-four Myles related this event:

> I thought about you the other day. I was very irritated with my supervisor and wrote an e-mail telling him about my complaints. I was about to press the send button when I remember what you say about shooting myself in the foot. I felt this might hurt my chances for promotion and I decided to erase it.

During the fifty-five months of phase two treatment not a single dream was presented. "Either I'm not dreaming or I can't remember them," Myles would say. Into this situation, in month fifty-six, the following dream was presented.

> I am in Calista's apartment or house. She lived above a restaurant or bar. There were only two large rooms. I felt there was a threat. Calista's daughter was there. Calista had or gave me a pistol. Calista killed Sabrina's sister, Dora.

Associations: Calista is the store director and as such is responsible for Myles' promotion.

> I felt strange in that I had just fallen asleep and thought how could I have had such a long dream so soon. I was really asleep only twenty or thirty minutes. Had Calista shot Sabrina I could have made sense of it. Is Calista trying to take me away from Sabrina in the job (by working me so many hours)? But Dora? Dora (Sabrina's older sister by 5 years) has had anger issues her whole life. Dora is a bartender (in real life), but in Calista's house! I didn't put Dora and Calista together. Is this something that gets in the way of my relationship with Sabrina? I have complained about Calista to Sabrina in very negative comments. Calista gives employees no recognition, is very critical of employees in public and, according to store gossip, her husband is a felon having served four to five years in a minimum security Federal facility in Indiana for financial fraud. Considering her behavior, I wonder if Calista is bipolar. Maybe Sabrina kills Calista. Calista was chewing out my supervisor on the floor yesterday.

The next session Myles said, "There was another part of the dream that I remember. My supervisor died at some point. Calista was in my department yesterday. We were standing by the printer and I accidentally stepped on her foot." "Now do you understand the dream?" I asked. "I knew you were going there," Myles said. "I want to kill her."

Discussion

The dream session felt like phase two treatment again. Myles felt a sense of achievement in arriving at a satisfying interpretation. The deletion of the potentially damaging e-mail clearly indicated that the therapeutic alliance is strong and that the therapist is present in Myles' psyche outside of the office. The overarching issues during this six-month period are stress management and the anxiety of dysregulation, which has a strong likelihood of being related to the stress he is experiencing and his 50% medication compliance.

In a consultation with the biological psychiatrist in the week following the dream session the following was ascertained: bipolar cycling seems to have a seasonal component with manic symptoms more prominent in the spring and depressive symptoms more evident in the winter. However, the shift to frank manic-depressive cycling is probably due to the drop-off in medication compliance. This drop-off may well also account for Myles' irregular sleep pattern as bipolar states affect sleep. Myles can stay off of Risperdal as long as the psychotic symptoms do not reappear. The psychotropic medications Myles is currently taking include Lithium Carbonate, melatonin, Cymbalta (duloxetine), Cogentin (benztropine), and now also Klonopin (clonazepam) to further enhance mood stabilization. The list of questions and concerns that Myles and I prepared for his visit to the psychiatrist were indeed discussed, leaving me with the feeling that the collaboration amongst patient, therapist and psychiatrist is synchronous and capable of yielding treatment synergies. The key selfobjects currently in Myles' life (Sabrina and parents) will be encouraged to support medication compliance.

The issue of the nature of how the term selfobjects is used needs some elaboration. See Chapter Five for a theoretical discussion of selfobject and supportive relationships.

Month 56 to month 62

Myles was developing the belief that his career was stalled; he was forming the conviction that the formal promotion would not materialize. Also, he was exhausted with the long hours and no end in sight to the seasonal relocation of inventory and reconfiguration of heavy merchandise displays, referred to as a floor reset. He began circulating his resume; we were practicing interview questions and working on his presentation. In month 58 he applied to a large chain of food stores. The human resources director was so impressed with his experience that he was interviewed two times in successive days and offered the position of department manager of the produce section pending satisfactory completion of a six-month probationary period. Myles gave the customary two-week notice to his current employer and left on a positive note. Myles was elated with this major career move and is now, at this writing, three months from completing the probationary period.

Attehdance at AA meetings has stopped and phone contact with his sponsor has diminished considerably. Myles has placed his first one-month sobriety coin along with his first voting sticker in a vase in the office. Sobriety is now a way of life supported by his entire selfobject milieu. The psychotropic medication now taken are lithium, Cogentin and Cymbalta, with Melatonin as needed for sleep. Myles is experiencing a challenging period of adjustment to a new work environment, different routines and coworkers, and expectations. His position is stressful in that as unsold produce ages it needs to be turned over and discarded if not sold by the freshness date. An obsessive-compulsive attention to detail is needed to be successful. This task is a core responsibility. To his surprise, an issue surfaced that was quite a jolt. In a meeting with his supervisor, HR director and the store manager, he was told in no uncertain terms that if he did not pay attention to his core responsibilities he would be terminated at the end of his probationary period. This issue came up, among others, in earlier meetings with these individuals, but he understood these interjections as friendly reminders. But now the gravity of it hit home. Myles said this issue followed him from his other place of employment. He was told repeatedly that his core responsibilities were getting neglected and his interests diverted to other projects. Instead of receiving compliments for his initiative and creativeness, he was again getting criticized. On the plus side, his supervisor and management team are perceived as empathic, supportive and want to see him succeed.

It was determined that the increase in goal-directed work activity, his elevated expansiveness, his creative enthusiasm and his boundless energy investment in unauthorized projects were classic symptoms of mania. He also remembers that when at restaurants with his girlfriend and/or mother he needs to be reminded to speak slower and more distinctly to the waitress as increased talkativeness and rapid speech is also a classic feature of mania (DSM-5, 2013, p. 124). The picture had become clear. Can the self-designated "Top dog in the new store" make the modifications and retain his promise of greatness and superiority? The focus now shifts clearly from mobilizing self-referencing metaphor to maintaining self-regulating metaphor. Can he exert the necessary willpower to stay focused on core responsibilities by using a strong force of concentration to pay attention to those intrusive, joyful emotions that disturb the work routine? In our sessions he works hard on recognizing these episodes and so far, for the most part, has demonstrated the resilience to stay focused. The motivation for success in this sphere of functioning is quite high. A sense of satisfaction has come to Myles now that he has awareness and understands that his central vulnerability is directly connected to his condition. Self-awareness of this vulnerability does not guarantee a perfect self-protective reaction. When he chases the tantalizing grandiose noncore projects he needs to know this is not an act of unconscious self-sabotage. The behavioral incentive is located in randomized biological brain processes, not unconscious acting out.

A second issue has occupied our therapy time – that of his body tremors. On occasion they can be quite noticeable. Should coworkers and/or supervisors inquire into his welfare and health, how would he respond? While frank disclosure of his condition is out of the question, Myles wonders if anything needs to be said. "Why do I have to say anything?" Myles said. After all it is not uncommon for employees to work satisfactorily with a host of medical conditions. Should he be pressed for an explanation, he was thinking about using intermittent fine motor control disorder, a neurological condition with more benign implications than the alternative full disclosure of his condition.

With the end of his three-month probationary period rapidly coming to a close, Myles now is receiving plaudits for his performance. Myles' functioning is now in alignment with his core responsibilities and the store manager's expectations. Recently Myles has said, "I feel like Chapman." This new self-referencing metaphor refers to the remarkable performance of Aroldis Chapman, ace closer for the Chicago Cubs in the 2016 World Series. We drew a distinction between performance as excellence versus grandiosity. Myles said, "I am in the strike zone. No reckless pitches."

Discussion

In this 13-month period, a great leap was taken with Myles leaving his five-year job. His decision was appropriate, well thought through and executed in a mature fashion. Now in his twenty-fourth year he has found an opportunity for a strengthened emotional stability. Demonstrating a sense of personal agency, Myles' employment relocation can be understood as a stress-reduction tactic. It is a way of dosing stress that offers opportunities for growth through limiting exposure to overwhelming stress that risks compromising the functional gains already made. Myles has acquired a 180-degree perspective on the impact of manic episodes on his functioning. Through a force of attention, he can manage these episodes, incorporating this technique into his treatment plan. With respect to the appearance of the fourth self-referencing metaphor, it is noted that all four are features of his functioning at points in time and suggest a continuity of progression of self-narration communicating his relationship to his selfobject milieu.

Next, I invite the reader into a series of reflections on the impact of two distinct types of child and adolescent rearing experiences on character formation and self-regulation. I present the Transformational Self concept, build a case for the superiority of a set of compatible interlocking theories, over a single theory, to explicate case dynamics, and introduce the developmental algorithm.

Note

1 The subject of alcohol use and abuse remains mixed in a fierce welter of complex and contradictory opinions on its origins, its benefits in moderation, the nature of individual and group vulnerabilities and motivations, and most of all, the efficacy of various treatment methods for alcoholism. While the gold standard for treatment appears to still be Alcoholics Anonymous, founded in 1935, at least four models vie for contemporary conceptual dominance: the biological/disease model (including AA), the moral model, the psychological model and the psychosocial model. (See attached chart in Appendix II.) Periodically, competent survey articles surface announcing that AA's leadership days are coming to an end (Delbanco, A. and Delbanco, T., March 20, 1995 "AA at the Crossroads." *The New Yorker*; Glaser, G., April 2015, "The False Gospel of Alcoholics Anonymous." *The Atlantic Monthly*).

While the controversies swirl about, it makes sense to me to understand the psychological significance of and value to the individual patient of the specific treatment. In the example of Myles, involvement in AA grounds him. He identifies with its message and methods. He values and relies on daily contact with his sponsor. Sobriety anniversary coins are highly regarded. AA serves a clear selfobject experience with vital functions in that it calms, regulates and stabilizes him. It is a bulwark against backsliding and regressing. Myles returns to his Intensive Outpatient Group to give testimony about how his life has turned around and so consolidates his identification with the AA model. So, while the universal efficacy of AA may be questioned, its value to certain patients, such as Myles, is beyond doubt.

4
ALTERNATIVE DEVELOPMENTAL MODEL THINKING

Lapham's wound

Every society has developed a prescription for the process by which a sensate child is transformed into a sensible adult. History suggests two fundamental pathways, one reasonably clear and the second ambiguous, to achieving adult status. They seem to involve either 1) participation in a group with a systematic indoctrination of the beliefs and values of that society and/or 2) an individual, personal struggle to acquire those beliefs and values.

Illustrative of the first pathway is public education in fifth century BCE Athens. By 420 BCE four levels of education had evolved. Girls, slaves and the poor were educated in an informal system. The four-level formal system, for boys only, began with elementary education at about age 7. In level two, post elementary education involved attendance at the gymnasium to age 14. From 14 to 17/18, boys could participate in a period of secondary education. In post-secondary education boys could move on to the fourth level, ephebic training. The ephebi (boys who have just become men) were 17- or 18- to 20-year-olds who formed a military cadet corps with garrison and patrol duties within the border of Attica on the way to becoming citizen-soldiers. They occupied a certain place of importance in public assemblies and festivals. After age 20 those with family connections and/or the most promise were selected for further education and preparation to assume greater public responsibility (http:/en.wikipedia.org/wiki/Education in Ancient Greece; Cambiano, 1995; *The Oxford Classical Dictionary*, 1996, pp. 528–529). To join the ephebi an oath was taken which, as a rite of passage, signaled with remarkable clarity, the parameters of the transition to young adulthood.

> I will not dishonor my sacred arms; I will not desert my fellow soldiers by whose side I shall be set; I will do battle for my religion and my country

whether aided or unaided; I will leave my country not less, but greater and more powerful, than she is when committed to me; I will reverently obey the citizens who shall act as judges; I will obey the ordinances which have been established, and which in time to come shall be established, by the national will; And whosoever would destroy or disobey these ordinances, I will not suffer him, but I will do battle for them whether aided or unaided; And I will honor the temples where my fathers worshiped; Of these things the gods are my witness.

(Bosanquet, 1990, pp. 8–9)

In contrast, consider the second type of pathway using Goethe's *The Sorrows of Young Werther* (Goethe, 1774/1971) as an exemplar. Goethe wrote *Young Werther* in six weeks at age 24; the story is considered loosely autobiographical. The plot line is as follows: Werther falls in love with Lotte who is unable to return his love because she is already engaged to Albert, who she soon marries. Werther forms a friendship with the couple, but is unable to accept Lotte's unrequited love and rejection and cannot move on. He falls into a deep depression and kills himself with his pistol, one that Albert had been holding for him. *Young Werther* became an immediate success and was popularized as an example of the "typical" *sturm und drang*, storm and stress, to be expected of adolescents. Young Werther initiated the fascination with the "necessary" struggle to attain adult status. Holden Caulfield (Salinger, 1945; Shields and Salerno, 2013) and Sylvia Plath (Plath, 1971; Claridge, Pryor and Watkins, 1998, pp. 200–211) are our contemporary versions of this struggle, a struggle sometimes with a psychotic core. Research has demonstrated that only about 25% of today's youth experience such developmental turbulence (Offer, 1980), but the myth of universal adolescent turbulence retains its hold on society. However, even if only one quarter of youth manifest psychological difficulty, all youth are exposed to our contemporary ambience of cultural destabilization awash in toxicity.

There is no doubt but that, for the moment, we have a generation of struggling adolescents. The number of students seeking counseling at university has risen by 50% in the last five years. The suicide rate is rising too, and the already scant mental health services cannot meet the ever-increasing need. In fact, the incidence of self-harm, depression, and suicide, especially among 14- to 15-year old girls, is currently the highest ever recorded. There are many elements that are recognized as contributing to this serious situation, although, none of them is, strictly, causal as such. They represent, rather, some of the reasons why adolescent angst about, for example, identity, gender, self-worth, exam performance, friendships, appearance – whatever recognizable issues there may be – are less caused by external factors alone than exacerbated by them. The external worlds that the young now inhabit are changing very rapidly, perhaps faster than ever before. They are having to contend with the internet, positive or negative, with social media, educational target-driven expectations, unemployment, and the rapidly changing possibilities and hazards of the digital age. All these can add considerably to already existent emotional disturbance. Massive technological and cultural changes have made possible some of the opportunities of the modern age, but also many of the horrors: cyber bullying, online sexual grooming and violation, virtual reality, gaming addiction, and innumerable other recent phenomenon – all deeply destabilizing – greatly add to the toxic mix of adolescent challenges … There is … no doubt but that socially, politically, psychologically, and ecologically, these are very troubled times. Media commentary, often focusing on the plight of young people, is laden with assertions that describe 'us' as, for example, undergoing an 'existential malaise', as being gripped by an 'epidemic of anxiety'. Catastrophic swathes of violence, war, terrorism, starvation, migration, famine, floods, despair, and death are sweeping across the world with a decreasing sense that anyone is in a position, or condition, adequately to control or respond to them.

(Waddell, 2018, p. 32)

In ancient Athens those vulnerable and developmentally at-risk ephebi were protected by group solidarity and supported by clear expectations and unambiguous behavioral parameters. By contrast, our technology-centered culture has no military draft and different state-by-state requirements and expectations of when to drive, to marry, to vote, to consume alcohol, to have an abortion, to sign a contract, to purchase marijuana and access firearms. Our heterogenous, pluralistic, multicultural fifty-state society values diversity with all its baggage of mixed messages and ambiguous expectations for the Millennial Generation. No wonder that our emotionally unstable Young Werthers, Holden Caulfields and Sylvia Plaths find themselves vulnerable and hard pressed to find those accommodations, including especially mental health services that can get them back on the developmental track and find a useful place in society.

As a contemporary society, we citizens remain preoccupied with how best to construct a purposeful, functional adult. Perhaps we always have. Brooks (2014), writing a recent op-ed piece, reflected on the purposes of a university education: the commercial purpose, starting a career; the cognitive purpose, acquiring information and learning how to think; and the moral purpose, building an integrated self. Quoting William James, Brooks writes, the morally significant life "is organized around a self-imposed, heroic ideal and is pursued through endurance, courage, fidelity and struggle" (p. A29). Can universities be expected to address the morally significant life? Deresiewicz (2014) recognizes that higher education excels at the commercial and cognitive purposes, but it has no idea how to build an integrated self. Citing Allan Bloom, Deresiewicz writes:

> True liberal education requires that the student's whole life be radically changed. Liberal education puts everything at risk and requires students who are able to risk everything. The process isn't comfortable, but it is exhilarating. There's nothing "academic" about it. If it happens right, it feels like being

broken open – like giving birth to yourself. "An education," Lapham quotes an old professor, "is a self-inflicted wound."

"Lapham's wound never heals, for the self that sustains it cannot return to innocent unconsciousness. What you should really want to develop in college is the habit of reflection, which means the capacity for change" (pp. 84–85). (Lewis H. Lapham is the former editor of *Harper's Magazine* and, since 2007, the editor of the literary magazine, *Lapham's Quarterly*.) The Lapham's wound metaphor is but the latest iteration of the *Bildungsroman* (coming of age) maturational theme.

The Transformational Self

So the powerful idea that one attains adulthood by going through a personal struggle leading to an individual transformation remains a persistent belief. Clinical experience suggests that for some adolescents a self-transformation is indeed possible and even desirable. I have formulated the Transformational Self (Bendicsen, 2013) concept to represent the manifold processes by which this transition occurs. It involves the spontaneous appearance of one or more highly personal, novel, self-referencing metaphors that have the potential of reorganizing the self, a reorganization that pulls the self forward with new identifications and a belief that there are latent capacities that can be harnessed. In other words, self-referencing metaphor evolves into self-regulating metaphor, becoming the pathway to personal metamorphosis. The metaphor(s) now serves as a north star channelizing the reconfigured self-state potentials into persistent behaviors that actualize and regulates the Transformational Self.

A necessary precondition for the emergence of the Transformational Self is the maturation of the prefrontal cortex with its enhanced neural connectivity. With this biological achievement in the late adolescent phase, executive functioning, a strengthened ego/self-state capacity, can arrive at a mature level of external stabilization and internal, intrapsychic structuralization. We can now speak about a neurobiological self, the neural self. The neural self is an integrated, enduring self-state that is subjectively experienced, has degrees of self-awareness and creates a narrative self within a single consciousness. Damasio (1994) fashioned the underpinnings to the neural self.

> The "neural self" is based on early body signals in evolution and development that helped form a basic concept of self; this basic concept provided the ground reference for whatever else happened to the organism. It includes current body states that are incorporated continuously and promptly become past states. The continuous reactivation of representations of key events in an individual's autobiography forms a notion of identity that can be reconstructed repeatedly in conjunction with activation of categorical memories that define an individual. The neural self consists of significant events in an individual's autobiography and body signals, which encompass past and

present background body and emotional states. "At each moment the state of the self is constructed from the ground up, it is an evanescent reference state so continuously and consistently reconstructed that the owner never knows it is being remade."

(Damasio in Greenman, April, 2007, pp. 52–53)

Schore (2002, pp. 443–448) contributed: "In addition, the neural self emerges as a result of right hemisphere maturation. It is a body self through the gradual differentiation of mutual co-regulation of state and affective experience between the caregiver and the self." Feinberg (2009) has defined the neural self "essentially by its coherence: a unity of consciousness in perception and action that persists in time" (p. xi).

As development unfolds, the neural self undergoes multiple and continuous reconfigurations to accommodate internal and external impingements. These impingements organize adaptation to specific states of differentiation. The phase-specific late adolescent neural self is rightly called a Transformational Self; it first emerges at the juncture between late adolescence and young adulthood. Also, different forms of transformational states can be apprehended at later significant developmental milestones or junctures.

The term "executive function" is one of the current high-profile cognitive/neuroscientific phrases that is subject to oversimplification. Spear (2010) helps clarify what exactly is meant by "executive function." "Careful assessment of cognitive function among adolescents, however, has revealed little evidence of either a single set of cognitive skills that would be characteristic of the formal operations stage, or any indication of a step-like shift in cognitive functioning from stage to stage" (p. 102).

Steinberg (2005) nicely summarizes this literature by saying that "rather than talking about a stage of cognitive activity characteristic of adolescence ... it is more accurate to depict these advanced reasoning capacities as skills that are employed by older children more often than by younger ones, by some adolescents more often than by others and by individuals when they are in certain situations (especially, familiar situations) more often than when they are in other situations."

(Spear, 2010, p. 102)

This description should remind the reader of the commanding role of complexity theory in human development. "Generally speaking, executive function can be viewed as 'higher-level' cognitive functions involved in the control and regulation of 'lower-level' cognitive processes and goal directed, future oriented behavior" (p. 105). Such lower-level cognitive functions usually, but not always, include: selective attention, goal setting, rule discovery, planning, decision making, response inhibition, working memory and cognitive flexibility. "*Development of more efficient executive and other cognitive functions sometimes is viewed as a shift from more diffuse to more regionally specific activation, particularly in frontal regions*" (p. 121). These regions include portions of the prefrontal cortex, parietal cortex and anterior cingulate cortex (Spear, 2010, p. 121).[1]

Let us now turn to the process by which reflection can be nurtured and promoted in a late adolescent to facilitate developmental movement. A set of interlocking theories will be used to formulate a theoretical understanding of the complex case of Myles. The instantiation of the Transformational Self, manifested in the appearance of a self-referencing metaphor, will illustrate its conceptual usefulness in building self-regulation.

Toward a compatible set of interlocking theories

Often the disciplines of neuroscience, developmental psychology and psychiatry function in isolation from each other. "Yet, when one attempts to synthesize their recent findings, an incredible convergence of many of these fields of study is revealed. These findings shed light on how the mind emerges from the substance of the brain as it is shaped by interpersonal relationships" (Siegel, 1999, p. 1).

> The ideas of this framework are organized around three fundamental principles:
> 1. The human mind emerges from patterns in the flow of energy and information within the brain and between brains.
> 2. The mind is created within the interaction of internal neuro-physiological processes and interpersonal experiences.
> 3. The structure and function of the developing brain are determined by how experiences, especially within interpersonal experiences, shape the genetically programmed maturation of the nervous system.
>
> (p. 2)

Siegel's three-part hypothesis is most useful in shaping the direction of inquiry into the contemporary integration of an interdisciplinary perspective on how mind, brain and interpersonal relationships mutually influence each other. Continuing, Siegel states, "This book attempts to synthesize concepts and findings from a range of scientific disciplines, including those studying attachment, child development, communication, complex systems, emotion, evolution, information processing, memory, narrative, and neurobiology" (p. 2).

Cozolino (2006) adopts a stance similar to that of Siegel. Cozolino takes the position that single brains cannot exist in isolation. "*The brain is an organ of adaptation* that builds its structures through interactions with others … *There are no single brains* … Interpersonal neurobiology assumes that the brain is a social organ that is built through experience" (p. 6).

> The social construction of the brain and the role of attachment relationships are particularly important in interpersonal neurobiology, as is the application

of scientific data to parenting, psychotherapy, and education. (Siegel & Hartzel, 2003). In addition to data from neuroscience and psychology, interpersonal neurobiology utilizes research from psychoanalysis, ethology, comparative anatomy, genetics, and evolution. In examining the social synapse we can look at narratives and storytelling, eye contact, touch, attachment patterns, and body language.

(p. 7)

A definition of regulation theory

Hill (2010) allows us to move to the last step and to stand on the landing looking out at a precise definition of regulation theory.

> Regulation Theory is an interdisciplinary approach to the body-brain-mind that integrates attachment theory, neurobiology, psychoanalysis, psychiatry, cognitive science, evolutionary biology, and infant developmental psychology. The theory provides an understanding of how mental states are organized around affect regulation, which involves the modulation of levels of arousal and maintenance of the organism in a homeostatic state in which the brain-mind can function optimally. The theory takes seriously that the mind, brain and body are mutually influencing subsystems of the organism. To function adaptively the organism must be regulated. The regulation of affect is the regulation of the organism. The capacity to regulate affect is developed in the attachment relationship to the caretaker. Patterns of affect regulation are activated each time one is involved in an attachment relationship.
>
> (pp. 80–81)

> The attachment function of the caretaker is to regulate the infant's affect, that is, the quality of the emotional tone and the level of arousal. Different patterns of affect regulation evolve as adaptations to the affect regulating patterns of caretakers. Such a pattern of affect regulation is imprinted onto the developing brain of the infant; that is, the regulatory style of the caretaker is internalized by the infant. Consistently sensitive caretaking generates secure attachment patterns in which affect is regulated and the brain-mind-body maintains a homeostatic state. Inconsistent or neglectful caretaking generates insecure attachment patterns in which affect is dysregulated and the organism functions suboptimally.
>
> (p. 81)

> The brain-mind-body requires regulation to function optimally. There is nothing more basic to the development of the organism. When the system is stressed beyond a certain point (dysregulated) it must be able to return to a regulated, homeostatic state where it can operate adaptively. There is a range constituting an optimal level of arousal, a "window of tolerance" in which

the system is regulated enough to remain flexible and stable. There is also optimal "response flexibility" to internal and external stimuli. In dysregulated states the system is either hypo- or hyperaroused and becomes organized into either a rigid or chaotic state. Response flexibility is lost under these circumstances, and one becomes overly dependent on the internal or external constraints on the system. Adaptive capacities suffer accordingly.

(p. 81)[2]

The point I wish to make is that Siegal, Cozolino and Hill use collections of domains of knowledge from various disciplines to advance their respective arguments and hypotheses, which, by the way, are in a good state of alignment with each other. I believe this is the contemporary intellectual trend, approaching the status of a *zeitgeist* (the general intellectual, moral, and cultural climate of an era, *Webster's New Collegiate Dictionary*, 1981, p. 1352), in applying and explicating neuroscience research to human development.

A definition of a developmental algorithm and the rationale for the selection of criteria in a developmental algorithm

I concur with the central hypotheses of interlocking disciplines of Siegel, Cozolino and Hill, but use a somewhat different mix of concepts, a variety on a theme if you will, to address the puzzle of how a late adolescent transitions into a young adult. Regulation theory requires an operational formula. The formula I espouse involves seven domains of knowledge linked into a metaphoric construct I label a developmental algorithm. I choose the mathematical term algorithm because of its definition of "a step-by-step procedure for solving a problem or accomplishing some end" (Webster's, 1981, p. 28). The end I seek is to establish the regulatory processes by which a late adolescent transforms into a young adult. My rationale for including the following seven domains of knowledge into my developmental algorithm are:

> **modern metaphor theory** because of its emphasis on embodied metaphor;
> **attachment theory** because of its emphasis on evolution and affect exchange as an early regulatory process;
> **self psychology with intersubjectivity theory and relational psychoanalysis** because of its emphasis on the nature of the self, the life-long need for selfobjects and the understanding of reality as a mutual, co-created intersubjective experience;
> **cognition** because of its emphasis on explicating individual thinking styles and learning patterns;
> **contemporary psychoanalytic developmental psychology** because of its emphasis on infant observation, evolution and non-linear processes;
> **complexity theory** because of its emphasis on open systems, attractors as dynamic systems organizers and continuous levels of organization and reorganization influenced by emergent and recursive system properties;

and **neuroscience with narrative theory** because of its joint emphasis on the continuous and discontinuous change of neural assemblies influencing ever-changing self-states reflected in the quality of coherence of autobiographical narrative.

I believe my developmental algorithm constitutes a core set of elements which, together, reveal an internal coherence. These seven elements exert a powerful synergistic compatibility that drives and expands explanatory power. Also, it is noted that frequently these respective domains of knowledge reference each other in the elaboration of their perspective on human development, thus enhancing functional explanatory compatibility.

The student of this subject may have observed that in my original developmental algorithm formulation I included relational theory (Bendicsen, 2013) as a coequal knowledge domain. After considerable reflection, relational theory has been repositioned in this iteration as a component of selfobject experience because, otherwise, its inclusion stretches the explanatory tent too thin. The deconstructed relational perspective does not comport well with attempts to organize the developmental trajectory. Nevertheless, intersubjectivity theory and relational theories remain vital and essential dimensions of human experience that must be taken into account.

There are two reasons for excluding relational theories as a stand-alone element from my developmental algorithm. First, relational theory, with its emphasis on here and now interaction, marginalizes the traditional psychoanalytic developmental perspective. Because a developmental algorithm is organically grounded in developmental model thinking, including the relational perspective seems to sound a note of conceptual incompatibility, weakening the linkage among the component elements. Second, the "self" as a unit of analysis and examination is, likewise, marginalized. Drawing from the psychoanalytic social constructivist philosophy and its contextual root metaphor, the "self" is understood as:

> social artifact that is shaped by the context in which the person is raised. The self is a historical phenomenon; it is not as an entity that has an essence and properties that endure through cross-cultural contexts. It is impossible to generalize about the self. There are no nomothetic (laws of nature, author's insert) propositions that have validity for the self.
> (Palombo, 1991, in Palombo, Bendicsen and Koch, 2009, pp. 355–356)

In my developmental scheme the best fit for relational theory is with self psychology because of the inherent intersubjective nature of the selfobject experience.

But relational theories do have a distinct perspective on the developmental experience. Writing from a "Process-Relational Paradigm," Witherington (2018) clarifies the relational perspective on human development, elaborating on the work of Overton (2015). Overton's work emphasizes developmental science as an interdisciplinary scientific field that lies at the intersection of psychology, biology, sociology, anthropology, cognition, neurobiology and social domains, among others. Eschewing the phrase "developmental model thinking," Witherington employs the term "developmental science." Process-relational developmental science joins three domains of knowledge, organicism or the organismic world view (Pepper, 1942), contextualism (Pepper, 1942) and dynamic systems theory (Thelen and Smith, 1994, pp. xiv–xvi). Organicism takes as its basic metaphor the belief than the organism must be understood from a system as a whole perspective with its movement viewed from its spatiotemporal particulars of activity-in-context patterns. Pattern explanations are not concerned with causes, antecedents or motivational forces. "Rather, pattern explanations are *atemporal, organizational* levels of explanation" that focus on organizational constancy, not the organism in the flow of time or the interaction amongst the system components (Witherington, 2018, pp. 43–44). Contextualism supplements organicism's neglect of activity in time by grounding itself in the here-in-now in the real-time dynamics of organismic activity. "Contextualism starts with the present event of specific action in a specific context and proceeds to other events immediately past and in the immediate future but never strays far from the spatiotemporal immediacy and context of the present" (p. 44) (see Palombo, Bendicsen and Koch, 2009, pp. xxxvi–xli for additional background). Dynamic systems theory reconciles "macroscopic invariant order at the level of systems as a whole ('the view from above') with microscopic variability and ceaseless fluidity of system parts and processes ('the view from below')" Witherington, 2018, p. 45). Dynamic systems theory "is unified behind a particular focus on the local-to-global, bottom-up dynamics of *emergence*: how macroscopic organization *arises* in open systems through the dynamic variability of real-time, local processes" (Witherington, 2015; 2018, p. 45).

The linkages among self psychology, intersubjectivity theory and relational psychoanalysis

In so far as intersubjective systems theory and relational psychoanalysis together constitute a paradigm shift, the distinction between them merits our attention. It is understood that neither intersubjective systems theory nor relational psychoanalysis has a developmental theory. I am not attempting to establish whether epistemology or ontology has priority. Joe Palombo has put it well: "I have avoided the duality between experience and being by embracing the notion that all experience is part of a process, even the 'self' is not considered an entity, but a way of being" (Private communication dated January 30, 2016). This discourse is an elaboration on the justification for including intersubjectivity and relational states (within theories) as dimensions of self-selfobject functionality in my developmental algorithm. Both are examples, with distinctions to be elaborated upon, of mutual co-created experience, intersubjectivity theory within the phenomenological tradition and relational theories within the postmodern philosophical social constructivist position.

Let us begin with intersubjectivity theory. Robert Stolorow and his collaborators (1978; 1979/1993; 1992) are central to the development of intersubjectivity theory. Impressed with Kohut's conceptual and methodological contributions, Stolorow et al. pushed beyond the ideas of self, object, selfobject developmental needs, selfobject transferences and the shift away from the central focus on interpretation to the sustained empathic immersion in the patients subjective reality.

> More importantly, "subjective world" is a construct that covers more experiential territory than "self." Therefore, an intersubjective field – a system formed by the reciprocal interplay between two (or more) subjective worlds – is broader and more inclusive than a self-selfobject relationship. An intersubjective field exists at a higher level of generality and thus can encompass dimensions of experience – such as trauma, conflict, defense, and resistance – other than the selfobject dimension.
>
> (Stolorow and Atwood, 1992, p. 4)

Drawing from object relations theory and infant research (e.g. Stern, 1985; 1988), the added emphasis included a more complete field or systems model. "Rather than the individual, isolated self, Stolorow's emphasis is on the fully contextual interaction of subjectivities with reciprocal, mutual influence" (Mitchell and Black, 1995, p. 167).

The linkage between Kohut's self psychology and Husserl's phenomenology has been articulated also by Reis (2011). Reis contends that in Husserl's first phase of thought, transcendental phenomenology, Husserl attempted to reveal the hidden essence of phenomena through interior meditation. Phenomena were to be bracketed from the external world and examined just as they are given to one's own consciousness. For some this approach suggested a withdrawal from "the intersubjective world for the private space of the solipsistic life" (p. 77). Husserl's second phase, that of transcendental intersubjectivity, influenced by Heidegger, marked a departure from the epistemic to the ontologic condition of life. "Husserl now viewed intersubjectivity as a being with others, and saw our capacity to constitute the world as intimately tied up with our already being in relation with other people" (p. 77). Knowledge is now understood as knowable only through one's own subjectivity and the common embodied subjectivities of the community of human-kind.

> Kohut's (1984) conception of the analyst as a participant-observer situated him or her within the relational field; and, if we understand Kohut to be employing a Husserlian conception of consciousness as intersubjectively open, then we may also say that Kohut's mode of empathy was an intersubjective knowing of the other, a disclosure of the other's experience in a shared life-world.
>
> (Reis, 2011, p. 79)

Hadley (2008) explains intersubjectivity as follows:

> Intersubjectivity is a developmental achievement, and one that most of us take for granted in day-to-day communication. It is a defining focus and feature in thinking and practicing from a relational perspective. Intersubjectivity is the capacity to recognize that others have their own subjectivity, and the ability to grasp certain aspects of what the other is feeling or thinking. This leads to the possibility of engaging in shared subjectivity, that is the basis of communication. Intersubjectivity enables us to consider the experience of each person (or subject) as he influences the experience of the other in ways that are reflexive, not unidirectional. The study of intersubjectivity has challenged us to broaden our understanding of development to include the emergence of the ability to recognize the other as a separate subject (Benjamin, 1995/1998) and to acknowledge the importance of a "mutuality of regulation, which refers to the reciprocal control that two people in a relationship continuously exert on each other" (Aron 1996). The relational approach to clinical work has been extended by theorizing about different "modes of subjectivity" (Mitchell 2000), and the renegotiation of the mutual impacts of client and therapist on each other to restore "the patient's sense of personal agency."
>
> (Hadley, 2008, p. 219)

The philosophical roots to intersubjectivity theory are found in the philosophy of phenomenology, specifically the works of Immanuel Kant (1724–1804); Edmund Husserl (1859–1938); Edith Stein (1891–1942), Husserl's first research assistant; Martin Heidegger (1889–1976), Husserl's collaborator and successor to Husserl's chair at the University of Freiburg; and in the writings of Hans-Georg Gadamer (1900–2002), who studied under both Husserl and Heidegger. This arbitrary line of development excludes the contributions of Franz Brentano (1838–1917) and Jean-Paul Sartre (1905–1980). Let us begin with Kant.

Kant's epistemology and metaphysics are found in his major first book, *The Critique of Pure Reason* (1781/1787). With perhaps the exception of Plato's *Republic*, it is considered the most important philosophical book ever written. Kant's philosophy is understood as an effort to synthesize the two very different philosophical positions of the day: rationalism and empiricism.

> Rationalists believe that the fundamental nature of things can be discovered by reason. According to the Principle of Sufficient Reason, everything must have a sufficient cause and in this sense: the universe behaves rationally according to principles grasped by reason. The window breaks. Given an adequate description of the causes, we see the window had to break. Given the causes, it was necessary that it had to happen in that way. Because of this, explanation will amount to logical demonstration. Effectively, this means that all truths are necessary truths, and it is in principle possible to know them

> through reason alone, without recourse to sense-experience. In any case, sense-experience is an inferior form of intellectual apprehension, according to the Rationalists, which does not give us knowledge of the cause of things.
>
> (Thompson, 2000, pp. 4–5)

Rationalism is grounded in Plato's belief, and later Descartes, that knowledge exists in forms of innate ideas in the human mind implanted at or before birth. Consequently, knowledge is acquired by helping educate people to become aware that these innate ideas exist. This position is represented by the *a priori*, deductive thinking of Gottfried W. Leibniz (1646–1716) (Muller, 1992).

> Empiricism of this period consists in two major principles. First, that all knowledge and concepts must be derived from sense experience. Second, that all we directly perceive are our own ideas. Together these principles have dramatically skeptical implications. Given them, how can we know things beyond our ideas, like material objects? How can we even have the concept of such things? Hume saw that given the Empiricist principles, many of our deeply held beliefs cannot be justified. Neither reason nor sense experience can justify beliefs in objects, causation and the self. The two Empiricist principles imply that there is no *a priori* knowledge of the world. According to Hume, all necessary truths are tautologies; they are analytic claims which merely reveal the logical relations between concepts and give no substantial knowledge about matters of fact. In this way, Empiricism denies that knowledge of the world can be gained from reason.
>
> (Thompson, 2000, pp. 4–5)

John Locke (1632–1704) rejected Plato's belief in innate ideas maintaining that there is no reliable evidence to support them. He became acquainted with empiricism and the scientific method of discovery and developed the idea that the mind is a blank tablet, tabula rasa, ready to receive sensations from the outside world and impressions from within. Knowledge is acquired in complex building block fashion with the mind analyzing and organizing data and discovering relationships among that data (Locke, *An Essay Concerning Human Understanding*, 1689; Leiser, 1992). David Hume (1711–1776) modified Lockean ideas about the knowledge of the nature of reality. Hume agreed with Locke that all our ideas are derived from our senses, but in particular, Hume took issue with Locke that the physical world and its independent existence outside of our awareness is knowable through our senses. Since our beliefs are based on imagination, they cannot be rationally justified. Hume believed that what we perceive through our senses are impressions that are not physical objects, but are contents of our consciousness. These contents emerge in our minds from unknown causes which are copies of our impressions. Because we cannot bridge the gap between the contents of our consciousness and an external, non-conscious physical world, we can have no knowledge of its existence. Because of our inability to have knowledge about the external world, the question of causation naturally follows. Hume believed that causation is merely an inward feeling of anticipation that is based on our past experience. Hume evolved a stance of complete skepticism with respect to causation and arrived at the inference that because our reality cannot be objectively known, all knowledge is subjective (Johnson, 1992; Rockmore, 2010).

Kant was powerfully influenced by both the rationalism of Leibnitz and Hume's empiricist epistemology. Kant's monumental synthesis began with the nature of experience: "our experience of an orderly world of objects results from the cooperation of two faculties – our senses and our minds. The first contribute the matter of our experience, the second contributes its form" (Johnson, 1982, p. 282). Our senses provide us with the content of the objective world of colors, smells, and so on, "But we do not gain our knowledge of the formal structure that experience of an orderly world requires through them" (p. 282). For example, there is no sensory experience of time or causality. It is the creative activity of the mind that organizes experience, not our senses. It is the mind that structures raw sensory data into coherent experiential structures. With this formulation Kant sidesteps the blank tablet mind of the empiricists, as well as changing the rationalists' concept of the mind from a substance (à la Leibniz, author's note) to an activity. To underscore the creative activity of the mind, Kant formulated the term *transcendental unity of apperception*. In this term Kant refers to three dynamics essentially. First, the word *transcendental* denotes the fact that the mind at work cannot be perceived through the senses. Its activity is a necessary condition of orderly experience. Second, the word *unity* refers to the mind's organization or synthesis of raw sense data to produce coherence. Third, the word *apperception* is distinguished from perception. Apperception is the placement of the element or object of examination into the totality of experience. Perception, on the other hand, yields only isolated sensations. With respect to the nonmental side of Kant's metaphysics, Kant's objective world is not that of our independently existing entity of physical reality. Rather, for Kant, "'objective' means a unified construct resulting from the combined activities of the senses and the mind. Kant's 'objects,' therefore, have no existence in independence of thinking observers" (Johnson, 1982, p. 283). Generalizing from Kant's account of "objective," a distinction can be drawn between the *phenomenal* world and the *noumenal* world, or the world of things-in-themselves. The phenomenal world refers to the world that we can experience, the world of sensory context. The noumenal world is the transcendental or external world, meaning a world falling outside our possible experience and knowledge, the "reality" of which we can only experience indirectly. Kant believed that his "objective" world, according to our ordinary beliefs about reality, is in fact a "subjective" world. "For this reason, Kant considered his philosophy a form of metaphysical idealism rather than realism" (Johnson, 1982, p. 285; see also Rockmore, 2010).

Husserl is considered the founder of the phenomenology movement in philosophy.

> Phenomenology is the study of structures of consciousness as experienced from the first-person point of view. The central structure of an experience is

its intentionality, its being directed toward something, as it is an experience of or about some object. An experience is directed toward an object by virtue of its content or meaning (which represents the object) together with appropriate enabling conditions.

Enabling conditions are those of possibility including context, intention, language, and so on. Further:

> Basically, phenomenology studies the structure of various types of experience ranging from perception, thought, memory, imagination, emotion, desire and volition to bodily awareness, embodied action, and social activity, including linguistic activity. The structure of these form of experience typically involves what Husserl called "intentionality," that is, the directedness of experience toward things in the world, the property of consciousness that is a consciousness of or about something. According to classical Husserlian phenomenology, our experience is directed toward – represents or "intends" – things only *through* particular concepts, thoughts, ideas, images, etc. These make up the meaning or content of a given experience, and are distinct from the things they present or mean.

Continuing:

> The basic intentional structure of consciousness, we find in reflection or analysis, involves further forms of experience. Thus phenomenology develops a complex account of temporal awareness (within the stream of consciousness), spatial awareness (notably in perception), attention (distinguishing focal and marginal or "horizontal" awareness), awareness of one's own experience (self-consciousness, in one sense), self-awareness (awareness-of-oneself), the self in different roles (as thinking, acting, etc.) embodied action (including kinesthetic awareness of one's movement), purpose or intention in action (more or less explicit), awareness of other persons (in empathy, intersubjectivity, collectivity), linguistic activity (involving meaning, communication, understanding others), social interaction (including collective action), and everyday activity in our surrounding life-world (in a particular culture).
> (Stanford, December 16, 2013, pp. 1–2)

Traditional phenomenology has focused on subjective, practical, and social conditions of experience. Recent philosophy of mind, however, has focused especially on the neural substrate of experience, on how conscious experience and mental representation or intentionality are grounded in brain activity. It remains a difficult question how much of these grounds of experience fall within the province of phenomenology as a discipline. Cultural conditions thus seem closer to our experience and to our familiar self-understanding than do electrochemical workings of our brain, much less our dependence on quantum-mechanical states of physical systems to which we may belong. The cautious thing to say is that phenomenology leads in some ways into at least some background conditions of our experience.

(pp. 1–2)

To the traditional methods of practicing phenomenology, neuroscience adds yet another.

In the experimental paradigm of cognitive neuroscience, we design empirical experiments that tend to confirm or refute aspects of experience (say, where a brain scan shows electrochemical activity in a specific region of the brain thought to subserve a type of vision or emotion or motor control). This style of "neurophenomenology" assumes that conscious experience is grounded in neural activity in embodied action in appropriate surroundings – mixing pure phenomenology with biological and physical science in a way that was not wholly congenial to traditional phenomenologists.

(p. 3)

In *The Cartesian Meditations* (1917/1950) Husserl clarifies his philosophical journey, phenomenology, as understood through conscious perception, as a descriptive science of direct experience. First, Husserl was not interested in empirical facts, but rather with the potential for what can possibly occur, with the essential structure of experience. Second, he was concerned with the way the mind constitutes experience, through the transcendental ego, the self-mechanism for the suspension of presuppositions, the foundation of reality. Each self constitutes a separate reality, including the reality of other selves. "However, each other self is intended as a self that in turn constitutes experience. The other is perceived through empathic pairing of an inner horizon to the bodily outer horizon." "Third, he was concerned with the way objects subjectively appear to human consciousness." For example, the contours of time became inexact and were understood as the holistic flow of "just-past" and anticipated "near-future." Husserl's intent was to add a dimension of "apodictic" knowledge or certainty to the study of intersubjectivity (Stiver, 1992, pp. 424–428). Husserl's phenomenology should not be confused with positivism and the direct observation of empiricism.

In Stein's *On the Problem of Empathy* (1917/1989) the nature of empathy is considered to be a central epistemological issue. While she presents arguments for empathy being an essential feature of our mental and psychological likeness to others, she attempts to differentiate and go beyond one's perceptions and the perceptions of others as constituting a coherent and comprehensive picture of reality. It is not adequate to infer that reality is known by analogy with our own case. She begins her inquiry by posing the question, "How can we know the minds of others?"

Stein's answer is that we know these things by *empathy*. Empathy is an irreducible intentional state in which both other persons and the mental states of other persons are *given* to us. In an empathic experience, we are *presented* with not mere bodies in motion, but rather with persons – and they are presented to us *as* persons who are angry, or who are grieving, or who filled with joy. Persons and their mental states are not theoretical posits or unobservable entities – they are objects of which we have something akin to perceptions.

(McDaniel, 2014, p. 3)

"Stein understands the problem of empathy as the problem of accounting for how other people and their experiences can be *given* to me despite their distinctness from me" (p. 1). Stein seems to be proposing that "the possession of empathy is a necessary condition for self-knowledge of one's material nature … not only must I have empathy, but there must also be other people who have empathy in order for me to have this kind of self-knowledge. Knowledge of my nature requires a community" (p. 14). Does Kant influence Stein's view of empathy? Agosta (2014) suggests a reconstruction of Kant's view of empathy as an *a priori* givenness of the other individual existing as the respect that one experiences in the presence of the other as an example of the moral law. (www.socialjusticesolutions.org/2015/11/01/empathy-kant)

Heidegger taught phenomenology for a decade, but when he took over Husserl's chair a serious disagreement arose over Heidegger's understanding of phenomenology as a method for ontology, the nature and relations of being. Husserl consistently and single-mindedly identified phenomenology with epistemology, while Heidegger consistently and single-mindedly identified phenomenology with ontology. Husserl and Heidegger had a progressively difficult and quarrelsome relationship with no hope of integrating or reconciling these two positions (Rockmore, 2010, pp. 141–143). Heidegger's highly acclaimed early work, *Being and Time* (1962), reoriented the understanding of being by creating a "fundamental ontology," a hermeneutic phenomenology. In this framework truth becomes a subjective dynamic due "in part to the fact that people approach everything with preunderstandings shaped by their situation in history." Consequently, all understanding is inherently interpretive or hermeneutic. In attempting to develop a language closer to the authentic experience of human existence he word-smithed a new philosophical vocabulary that became part of the lexicon of the existentialist movement. (Stiver, 1992, pp. 519–523).

Maurice Merleau-Ponty (1908–1961) was a French phenomenological philosopher, strongly influenced by Husserl and Heidegger. He

emphasized the body as the primary site of knowing the world, a corrective to the long philosophical tradition of placing consciousness as the source of knowledge, and that the body and that which it perceived could not be disentangled from each other. The articulation of the primacy of embodiment led him away from phenomenology towards what he called 'indirect ontology' or the ontology of the 'flesh of the world' ... seen in his last incomplete work, The Visible and Invisible.

(Wikipedia, Maurice Merleau-Ponty, November 4, 2015, p. 1)

Put differently,

the flesh is neither some sort of ethereal matter nor is it a life force that runs through everything. Rather it is a notion which is formed in order to express the intertwining of the sensate and the sensible, their intertwining and reversibility. It is this notion of reversibility that most directly problematizes the concept of intentionality, since rather than having the model of act and object, one has the image of a fold, and of the body as the place of this fold by which the sensible reveals itself.

(Stanford, June 14, 2004, p. 14)

Additional philosophical roots are found in the work of Friedrich Schleiermacher (1768–1834) and Wilhelm Dilthey (1833–1911) and their work on establishing the foundations of modern hermeneutics as a field of major inquiry in the humanities. Hans-Georg Gadamer (1900–2002) extended hermeneutics to include not only textual explanation, understanding based on formulating meanings and framing interpretation through the inner life of the author, but establishing a philosophical understanding of hermeneutics as an ontological process of man in which understanding is constantly being recreated anew (Stern, 2013; Grondin, 2004).

Let us now move to relational theory. Stephen Mitchell (1988, and Black 1995, 2000) and others such as Stern (1997), Davies (1994), Bromberg (1998), Bollas (1987), Hoffman (1983, 1988), Aron (1991; 1993; 1996; 2005), Maroda (1991), Harris (1991; 2005) and Benjamin (1988; 1995; 1998) are considered among the seminal contributors in the founding of the relational school of psychoanalysis, perhaps the newest of the psychoanalytic theories. Relational theory, which actually is many theories, is understood by some as comfortably aligned with intersubjectivity theory, but its philosophical roots are different. Relational theory lies squarely within postmodernism, the philosophy of social constructivism. Its heritage traces back to the work of Harry Stack Sullivan and the interpersonal school. Sullivan was influenced by George Herbert Mead (1913) in their days together at the University of Chicago and Mead's work on the nature of the self, empathy and intersubjectivity (Cottrell, 1978).

The relational school represents a co-constructed meaning-making experience between analyst and analysand in the context of an extreme relativism. Truth is contextualized. In relational theory, theoretical universals are abandoned, knowledge is deconstructed. The two perspectives share an epistemology called "perspectival realism" defined as

Each participant in the inquiry has a perspective that gives access to a part or an aspect of reality. An infinite – or at least an indefinite – number of such

perspectives is possible ... Since none of us can entirely escape the confines of our personal perspective, our view of truth is necessarily partial, but conversation can increase our access to the whole.
(Stolorow, et. al., 2002, pp. 109–110 in Ringstrom, 2010, p. 198)

As I understand it, social constructivism is a neo-Kantian view of ontology and epistemology. Epistemologically, the external world is not knowable. Ontologically, it is structured by the categories of the mind, hence it is "constructed." Relational theorists then conclude that the categories through which we understand the world are based on the social context that we inhabit. That is the root of their relativism. Every society transmits to its members ways of seeing the world, categorizing its contents, and shaping its member's experiences. This view is worlds apart from the phenomenologists.
(Personal communication from Joe Palombo, September, 6, 2015; Johnson, 1992, pp. 281–285)

Intersubjectivists tend to be more bidirectional, relationalists more multidirectional with more of a blurring of the distinction between subject and object. The relationalists emphasize deconstructed experience in a global multicultural context.

Mitchell describes the relational perspective:

In this vision the basic unit of study is not the individual as a separate entity whose desires clash with an external reality, but an interactional field within which the individual arises and struggles to make contact and to articulate himself. Desire is experienced always in the context of relatedness and it is that context which defines its meaning. Mind is composed of relational configurations. The person is comprehensible only in the tapestry of relationships, past and present. Analytic inquiry entails a participation in, and an observation, uncovering, and transformation of, these relationships and their internal representations.
(Mitchell, 1988, p. 3 in Hadley, 2008, pp. 210–211)

To summarize, the similarities between intersubjective systems theory and relational psychoanalysis are that both deemphasize drive theory, turn away from a linear developmental view, find a more compatible fit with non-linear dynamic systems theory, represent variants of systems/field theory and share an epistemology.

Among the differences are:

1) Intersubjective systems theory is influenced by the British middle object relations school, self psychology, attachment theory, infant research and affect regulation theory. Relational psychoanalysis, in addition, is influenced by the interpersonal school, Kleinian psychoanalysis, feminism, gender theory, work with oppressed populations, and cultural diversity.

2) Following Ganzer (Personal communication of October, 14, 2015), Ringstrom (2010) points out that intersubjectivists tend to write in one unified voice, while relational authors form a collection of disparate voices. This multiplicity makes it difficult to characterize the relational model as a single, coherent theory.
3) In the intersubjective model, the patient receives more attention whereas the relational model gives more or less equal status to both participants, and, at times, pays more attention to the therapist. Some researchers believe this is what makes the relational model different from other contemporary theories.
4) The intersubjectivists formulate the intersubjective matrix as an abstract field upon which all experience occurs. It is the holistic perspective of Gestalt theory, including the totality of the environment. The relationalists view intersubjectivity as a developmental process of capacity, à la Benjamin (1990/1995). Here capacity refers to a process of increasing ability to discriminate subject from object, to moving "beyond viewing the other as an omnipotently controllable (or controlling) object to an awareness of the other as an irreducible subject of initiative in his own right" (Ringstrom, 2010, p. 201). The intersubjectivists use the term "intersubjective-field" and relationalists use the term "relational matrix" or "relational field."
5) The intersubjectivists use empathy as the primary mode of understanding patients. Whereas the intersubjectivists are more focused on immersion in the patient's experience, the relationalists are more perspectival and look at empathy also from the otherness of the therapist.
6) With respect to the patient's vulnerability, intersubjectivists and relationalists occupy different ends of a continuum. For the intersubjectivist "the therapist must be extraordinarily vigilant, so as to not provoke 'pathological accommodation' to the analyst's narcissistic needs." For the relationalist, "there is concern for over the analyst's vulnerability to 'pathologically accommodating' the patient's view in a matter that fortifies her patient's omnipotence, further isolating him from the world." These positions counterbalance each other (pp. 204–205).
7) "The most significant difference between intersubjective and relational practice is the use of projective identification. Intersubjectivity, with its roots in self psychology, has empathy and attunement as primary foci. Although not particularly influenced by the Kleinians, relational theory has incorporated projective identification as a means to signal enactments" (Ganzer, op. cit.).

Wolf (1988) has expanded the utility of the selfobject concept by emphasizing the bidirectional nature of efficacy experiences "characterized by the self as the actor and the selfobject as the acted-upon" (p. 60). His definition of selfobject indubitably places selfobject functionality in the intersubjectivity/relationalist orientation.

> A selfobject is neither self nor object, but the *subjective* aspect of a self-sustaining function performed by a relationship of self to objects who by their presence or

activity evoke and maintain the experience of selfhood. As such, the selfobject relationship refers to an intrapsychic experience and does not describe the interpersonal relationship between the self and other objects. It denotes the experience of imagoes that are needed for the sustenance of the self.

(p. 184)

I understand that as I discuss the formulation of the concept of a "developmental algorithm" in functional terms I run the risk of continuing the tradition of objectification of psychoanalytic concepts through the use of metaphoric devices. Self psychologists are well aware of this tendency (Wolf, 1988, pp. 11, 27–28). Selfobject functions and self-experiences are amply documented (Palombo, Bendicsen and Koch, 2009, pp. 263–265). Another example moves us into the blending of concepts from executive functionality, self-deficits and neuroscience.

In his discussion of "executive function," Goldberg (2001, p. 23) compares the frontal lobes of a human being to the CEO of a corporation or an orchestra's conductor. Its functions involve the capacity to impose order among activities, to plan future actions and to organize the sequence in which those actions should unfold. He suggests that among the talents of those who possess good executive functions are the capacity to be "smart" and "shrewd."

(In Palombo, 2011, p. 282)

However, Summers (2013) entreats us to recognize the incompatibility between experience as intersubjectivity and experience as reification of metaphoric conceptualization: "the development of the science of human experiencing requires the elimination of spacialized concepts so often used in in common discourse as well as in psychoanalytic theory" (p. 182). Summers continues:

I know who I am because I experience myself as the same at different points of time, and others know me by connecting their experience of me temporally. The temporality of being matters for analytic theory because it reformulates the nature of the self and opens up the temporal for exploration in a new way. Temporality cannot be "spatialized," so it lends itself to experience and offers protection against reification.

(p. 183)[3]

Nevertheless, it seems to me that as a hermeneutic device, a case can be built for describing an experience in functional terminology and at the same time experiencing it subjectively. If we stop discussing the self in spatial terms and move to reflecting on the self as located only in time, we will need a bridging concept because we indeed do live in and along a space-time continuum. All of us are located in both dimensions of experience. At this point in my labors I may have to accept the potential for incompatibility in my theorizing and schedule the quest for a more rigorous conceptual clarity and constancy as a goal for a future project.

Theoretical considerations

Let us spend some time thinking about the perspectives all clinicians face when deciding how to organize data into a diagnostic formulation. What do we do with the theories that powerfully shape our formulations? Our theoretical formulations are informed customarily by either: a single theory, a pluralistic array of theories, eclecticism or an attempt at an integration of theories or an avoidance of integration as in "epigenetic hierarchical arrangement" (Jaffe, 2000). I want to move to an examination of each of these formulations, but first some context.

Before proceeding it is efficacious to position this developmental algorithm in relation to the dominant philosophical perspectives orienting contemporary clinical and developmental thinking – the positivist and the social constructivist positions.

From a social perspective, commentators in both public and private spheres comfortably divide the world into "hard" and "soft" ideological positions.

> Computational theory of mind falls into the hard-thinking camp. Thoughts are not literally hard or soft, of course, but the associative train is easy to follow. Clarity, precision, sharp outlines, logic, and dispassionate objectivity belong to hard thinking and hard science, while fuzzy, imprecise, mushy borders belong to soft-thinking poets, novelists, scholars in the humanities, and other subjective and emotional folk.
> (Hustvedt, 2016, p. 363)

From a historical view, scholarly philosophical inquiry traces this division to two basic sources. The first source is Descartes (1596–1650), a French philosopher noted for his all-consuming search for certainty and a "legacy of a dualism of two distinct and independent created substances, minds, which were unextended in space and were thinking, and material bodies, which were extended in space and could not think." Descartes' mental substance and his material body appear to interact causally, but this interaction is left ambiguous (Clarke, 1992, p. 199). Quoting Hustvedt:

> The subject/object problem is an old and fraught one, and many a philosophy student can recite the narrative, which may be said to begin with Descartes's extreme doubt about knowing and himself, which is exploded by Hume, then reconfigured by Kant in his answer to Hume, and after that returns in a new form in German idealism and is further transformed in phenomenology with its supreme focus on the first person in the work of Husserl, who influenced Maurice Merleau-Ponty but also Heidegger, who then had considerable influence on poststructuralism, posthumanist continental thinking.
> (Hustvedt, 2016, p. 347)

Hustvedt suggests a second source for this division, that of Giambattista Vico (1668–1744), an Italian educator and the first philosopher of history. Vico "believed that the mind changed in history and that forms of thought, although influenced by the body, are not inflexible. In Vico, the Stone age mind becomes a modern mind" (p. 298)[4]. Vico developed a knowledge framework replacing Descartes' *a priori* concepts as the basis for science and introduced his concept of truth, *verumfactum* (truth is what is made or done). Truth is distinguished from reality without any necessary connection between the two. In agreement with Descartes that self-knowledge derived from introspection is the most profound knowledge, Vito introduced an ordering of knowledge based on degrees of certainty. In so doing he eventually become known as the anti-Descartes (Moss, 1992, pp. 248–249). "While Descartes discovered truths that were static and universal, Vico's truths included language and historical change" (Hustvedt, 2016, p. 147). The profound philosophical legacies of Descartes' and Vico's thinking are described, respectively, as follows:

> Positivists contend that science is a systematic public enterprise controlled by logic and empirical fact, whose purpose is to formulate the truth about the natural world (see Bernstein, 1983). Sensory observation is the source of external or experience distant data. Self-reports from patients of their introspections, which Freud believed to be obtained through evenly hovering attention and association, have their source in the internal near psychological events. Both of these sources yield equally valid data. Natural laws emerge from these observations and reflect an order inherent in nature. These laws or general hypotheses may be ordered into a hierarchy of increasing generality and complexity. Testing these hypotheses involves an appeal to facts disclosed in common observation of data. Predictions are possible based on tested hypotheses. The vision is of a universe of objects with independent existence (see Scheffler, 1982).
>
> (Palombo, Bendicsen and Koch, 2009, p. xli)

Critics of positivism, broadly defined as the postmodern movement, such as intersubjectivists and relational theorists, offer alternate views. They hold that realities are multiple rather than singular and fixed. All data are theory bound and contextual rather than objective and decontextualized; the observer and the observed cannot be separated. Since it is not possible to establish causal relationships between events, only the recognition of patterns of sequences of events is possible, and finally, inquiry is never entirely value free (Guba, 1990). These principles lead to the conclusion that theories are ideographic, that is they provide descriptive accounts of the patterns to which the phenomenon they describe conform. Each discipline bases itself on different belief systems, different methodologies and each aspires to different goals. Some radical critics of positivism go so far as to claim that even the natural sciences offer no more than sophisticated culture bound theories of the segment of the universe that explain. Others insist that there are irreconcilable differences between the natural and the social sciences. They claim that while positivist approaches are successful for the

natural sciences, constructivist or hermeneutic approaches are more appropriate for the social sciences (Saleeby, 1994).
(Palombo, Bendicsen and Koch, 2009, pp. xli–xlii)

Let us consider the issue of integration of data, or rather the integration of perspectives.

At this point, most informed theorists agree that the goal of integrating perspectives from empiricism and hermeneutics is beyond our reach. But first, what is meant by "integration"? *Webster's Dictionary* (1981, p. 595) lists the following definitions of integration according to four perspectives:

a) Culture: incorporation as equals into society or an organization of individuals of different groups or races;
b) Mathematics: the operation of finding a function whose differential is known or the operation of solving a differential equation;
c) Medical: the combining and coordination of separate parts or elements into a unified whole as the coordination of mental processes into a normal effective personality or with the individual's environment, or the process by which the different parts of an organism are made a function and structural whole especially through the activity of the nervous system or the hormones;
d) Legal: a writing that embodies a complete and final agreement between parties.

Taking these various perspectives into account, I propose the following definition of integration of theoretical orientations as the process by which 1) different systems of thought are combined into a unified whole, 2) such that fundamentally different elements are brought into functional, explanatory relationships with each other, 3) with the process measured by the degree of coherence. Examples of this definition of integration can be found biologically in the relationship between the sympathetic and the parasympathetic divisions of the central nervous system (Tortora, 1986, pp. 484–485) and psychoanalytically in Freud's description of the relationship amongst the mental systems Cs, Pcs and Ucs (Freud, 1915).

At present there are two dominant philosophical groups vying for explanatory dominion – the post-positivists and the social constructivists. Such a unified whole is possible when a theory is conceptually located into either a positivist or social constructivist camp. Palombo may have attained such an achievement with his neuropsychodynamic theory (2017b) located entirely within the positivist camp of scientific realism. Palombo maintains that to mix elements and descriptive metaphors from both camps runs the risk of conceptual incoherence. However, such a risk can be minimized if care is taken to employ concepts that complement each other without undue emphasis on integration. A developmental algorithm finds its home in pluralism, bypassing the need for bridging concepts in the endless search for seamless unity. A developmental algorithm is neither eclectic nor integrative, stressing rather the search for optimal coherent explanatory synergy.

A word about single theory case formulation is in order here. Usually single theory formulations are most helpful serving to clarify dynamics associated with a focal developmental issue or a specific psychopathology. Freud's *Little Hans* (*Analysis of a Phobia in a Five-Year Old Boy*, 1909) comes to mind as an exemplar illustrating specifically the developmental vicissitudes of the oedipus and a childhood neurosis. Freud drew inferences from *Three Essays on the Theory of Sexuality* (1905) and, what has come to be known as, the psychosexual framework, to validate his speculations grounded in his drive/defense theoretical model. Other examples of single theory formulations from different theoretical perspectives can be found in *Case Studies in Psychotherapy* (2005). There are at least two limitations to single theory formulations: 1) they explicate a specific psychopathology and/or a focal developmental issue and so may explain less well other psychopathologies or focal developmental issues; 2) they sometimes have significant autobiographical dimensions that introduce a bias and so can compromise objectivity. While the bias does not invalidate the explanatory insight, it makes it difficult to claim universality. See *Guide to Psychoanalytic Developmental Theories* (2009) for examples of autobiographical influences in theory construction. Other case formulation approaches, such as compatible interlocking theories as in a developmental algorithm, allow for a larger theoretical tent. Judging from a sampling of the recent books printed from psychodynamic publishers, the current intellectual trend, or *Zeitgeist*, appears to be one of harnessing a combination of theories to multiply explanatory impact.

Before proceeding, some elaboration on the clarification between eclecticism and pluralism is necessary. Rangell (2002) regarded pluralism as a potential unitary theory, an antidote to the theoretical state of disarray psychoanalysis finds itself in after its first century.

> In the ongoing debate over "one theory or many" (Rangell, 1988, 1997), I have advanced the idea of a "total composite psychoanalytic theory." This theory is unified and cumulative. It is total because it contains all nonexpendable elements; composite because it is a blend of the old and all valid new concepts and discoveries; and psychoanalytic as fulfilling the criteria for what is psychoanalysis. Every viable contribution made by alternative theories finds a home within this total composite theory. Accommodated under its embracing umbrella are drives and defense, id, ego and superego, self and object, the intrapsychic and interpersonal, the internal and external worlds. This theory aims for both completeness and parsimony. Such a unitary theory is not monolithic. Within it are many principles of multiplicity, such as Freud's theory of overdetermination, Waelder's principle of multiple function (1936), and Freud's multiple metapsychological points of view converging on any single psychological phenomenon.
>
> (Rangell, 1988, p. 1126)

Rangell wisely cautions us against the *pars pro toto* mechanism in which a form of thinking selects a part and substitutes it for the whole (p. 1115). If the theoretical part is forceful enough, its theoretical disciples attempt to develop it into a superordinate theory, thus fostering division and divisiveness. Rangell's pluralism, while compelling, advocates an "overlapping center" of conceptual harmony, grounded in drive theory, leaving out or deemphasizing nonpsychoanalytic interdisciplinary contributions.

A better way to think about pluralism is found in the work of Samuels (1989). Reflecting on the destructive schisms in psychoanalytic theory, he maintains that the pluralistic position, as a tool or instrument, can embrace theoretical and political diversity as a counterweight to schismatic tendencies. Pluralism seeks to embrace a perspective in which diversity, not unity of schools, is the goal. Rather, the goal privileges a modular approach in which different worldviews meet, but do not try to take over each other. In the absence of a hierarchy of theories, cooperation, not competition among theories, is emphasized. The challenge of cooperation is both intensely emotional and passionately ideological. It recognizes that there is an inherent interdependence among diverse theories with all manner of convergence and divergence of perspectives. Pluralism promotes an attitude toward dispute and disagreement that has been termed "bootstrapping."

> This somewhat unfortunate term is borrowed from theoretical physics. Instead of searching for one guiding theory, we might consider using many theories in parallel up to a point where their mutual inconsistencies and incompatibilities cause this to break down. Then the breakdown becomes a focus for study. Bootstrapping is not the same as synthesizing theories nor is it a form of eclecticism based on consensus. What is central to bootstrapping is that no one theory, nor the level of reality to which it refers, is regarded as more fundamental than any other. Geofrey Chew, the originator of the bootstrap approach, writes: "A physicist (read analyst) who is able to view any number of different, partially successful models without favoritism is automatically a bootstrap-per." The bootstrapper knows that it may be impossible for there to be a single theory which will work. What holds the theories together is that the subject matter somehow holds together: for the physicist, the universe; for the depth psychologist, the psyche. Passion for one approach, inevitably partial, is replaced by a passion for a plurality of approaches.
> (Samuels, 1989, p. 39)

Pluralism is not eclecticism. Eclecticism extracts parts of theories from their respective whole and ignores the boundaries and contradictions between systems of thought. A developmental algorithm is an example of Samuel's form of pluralism – all seven bootstrapped components contribute toward a coherent explanation, each in their own medium of origin and matrix of complementarity.[5]

A recent arrival to this exercise about applying different theories to explain clinical data sets is that of Jaffe's "epigenetic hierarchical arrangement" adapted from Gedo's (1988) classification system. Recognizing that the five core psychoanalytical orientations, drive, ego, self, object and relational, exist as silos, if you will, of separate, incompatible theories, spending time and effort finding commonalities has not produced a satisfactory clinical yield. Rather than focus on attempting to integrate these core theories, Jaffe writes,

> I do however see the yield of clinical observations and inference from treatment process as classifications of typical developmental configurations, but without reference to a core driving motivation. Put another way, I think the clinical formulations (Oedipus, transference, particular configurations of object relations, self-stabilizing experiences, etc) are useful descriptions but lack causal explanatory value. So the theories don't require integration because these descriptions (classifications) stand on their own and can be grouped roughly in a developmental array (I use Gedo's epigenetic hierarchical arrangement). They are connected to the rest of the scientific world by the elements present in complex systems that do not require instincts, drives, egos, selves, or egos as primary movers of development and function.
> (Personal communication February 5, 2018)

Jaffe's quest is to create a biological foundation for psychological development and therapeutic change (Jaffe, 2000). Elaboration of this foundation is eagerly anticipated (Personal communication with Charles M. Jaffe on February 5, 2018).

Let us now move to an exercise in which a developmental algorithm is applied to the case of Myles. This exercise is followed by the presentation of a framework of developmental forces and processes (or vicissitudes as in change or variation, if you will) that are generally understood to propel adolescents into young adulthood. This dynamic framework is presented as an alternative to the more static "tasks to be addressed" traditional model (Blos, 1962; Colarusso, 1992) generally used to characterize this developmental phase.

Perhaps a clarification of the word "algorithm" is in order. With the definition of algorithm being "a step-by-step procedure for solving a mathematical problem or accomplishing some end" (Webster's, 1981, p. 28), some students have suggested that the term algorithm connotes mathematical exactness resulting in producing concepts and principles with the certitude of dogma. Rather, my intention in using the word is to imply mathematical probability (as in the difference between inference and conclusion) with the promise of possibility and the potential of for example, hypothesis formulation and testing. The term "developmental algorithm" is designed and employed to contribute toward solving a developmental problem first articulated in *The Transformational Self* (2013), "When does adolescence end?"

Notes

1 Newspaper reporters, Belsha and Girardi, reported recently on a terrible auto accident outside of Chicago. A seventeen-year-old boy was driving an SUV carrying six students on their way to a regional basketball tournament game when it collided with a semi. Two sixteen-year-old students, one boy and one girl, were killed. The driver was seriously injured and taken to a hospital; the other three sustained minor injuries and were released

at the scene after signing medical wavers. All students, except one, were wearing seat belts. The evening was warm and foggy resulting in poor visibility. The SUV ran through a red light colliding in "t bone" fashion with a semi. It appears to have been "a joyful mission to support their school in a playoff basketball game" There was no report of alcohol or drug involvement. It seems that the 2007 Illinois revised teen driving law was complied with in that the driver was not cited for too many passengers. How can this accident be understood? (Belsha and Girardi, 2015, pp. 1 and 11).

One way to think about this all too common driving tragedy is to employ findings from neuroscience-related research, viz., executive functioning dynamics. First, cognitive studies suggest that multiple distractions, such as cell phone usage, listening to the radio/CD player and conversations in a confined space, along with the normal complexities of driving, especially, in bad weather, place the teen driver at great risk. The immature brain is unable to multi-task well and is susceptible to sensory overload. Second, the immature adolescent brain is supersensitive to rewards that makes them "naturally more attentive to the good things that might arise from their risky behavior." The brain's reward center is a strong motivational factor, stronger than the adult reward center that can override good judgment. Third, imagine the joyful peer companionship and camaraderie associated with the anticipation of an exciting event.

> The social brain is still changing in adolescence, and these changes help explain why young people's concerns about what their peers think increase during this time. It's the perfect neurobiological storm, at least if you'd like to make someone painfully self-conscious: improvements in brain functioning in areas important for figuring out what other people are thinking, the heightened arousal of regions that are sensitive to social acceptance and social rejection, and the greater responsiveness to other people's emotional cues, like facial expression. Given all of this, it is easy to see why changes in these parts of the brain increase adolescents' sensitivity to their status within their peer group, make them more susceptible to peer pressure, and make them more interested in gossiping (and more anxious about becoming the subject of other people's gossip). Brain scientists have uncovered the neurobiological underpinning of all that social drama.
>
> (Steinberg, 2014, p. 95)

The convergence of one or more of these factors result in a "hot cognition" and a loss of focus on the primary task – safe driving. (See the social brain section, *viz.*, the vagal system of autonomic regulation in this volume.)

2 The concept of regulation theory or adaptation has accompanied the discipline of psychology from its beginnings. Consider these passages:

> To put it simply, the equilibrium principle asserts a relation between a system (or an organism) and the environment in which it functions, such that if the environment changes, the organism will adjust its behavior to maintain certain desired conditions. Should the environment become modified, the system (be it living or nonliving) will vary its behavior so as to preserve itself.
>
> (Haroutunian, 1983, p. 2)

And

> Here then is one application of the equilibrium principle. Spencer uses it as a basis for explaining the occurrence of mental states, arguing that if the conditions in the organism – the relation between mental states *a* and *b* – are not consistent with the relation between corresponding external elements, then equilibrium between the organism and the environment is upset. In this situation, the relation

between the mental states must be modified if desired internal conditions are to be reached and equilibrium is to be reestablished.

(Spencer, 1892, in Haroutunian, p. 3)

3 It is well known that Freud adopted both scientific and hermeneutic approaches to observing the human condition. The emergence of neuropsychoanalysis extends the first approach. Beginning with Dilthey (1900), the hermeneutic and intersubjective schools have gained in theoretical popularity. Summers (2013) vigorously continues this tradition along with others, notably Hoffman (2002).

4 During our evolutionary history sensate mankind gradually harnessed language which enabled the channeling of fantasy life into a series of global metaphoric, if initially unstable intellectual constructs. These constructs, bridging objective and subjective experience, created a sense of agency or perceived control over our tenuous existence through the power of compelling explanatory theories or hypotheses. The basic hypothesis, called a root metaphor, helped organize life by injecting a measure of certitude and predictability, and creating a bulwark against nature's capriciousness. Root metaphors provided a sense of intellectual coherence in modeling the universe and re-describing parts of our experience in the world. Pepper (1942) posited six root metaphors.

> They are (a) *animism*, the notion that all nature is imbued with life; (b) *formism*, the Aristotelian concept that each organism has within it the seed of its structure, which will guide its development; (c) *mysticism*, the belief that a person may merge with nature or the universe to attain a higher level of being; (d) *mechanism*, the concept that all processes including those of human development may be understood as analogous to a machine; (e) *organicism*, the theory that all living matter, as organisms, may grow through the ingestion of nutriments, and follow a developmental sequence; and finally, (f) *contextualism*, which is the view that the best approach to understand all human phenomenon is to view them in their historical contextual environment and understanding their meaning.
>
> (In Palombo, Bendicsen and Koch, 2013BIB-030, pp. xxxvi–xxxviii)

The understanding of this sequence is that while succeeding metaphors may hold conceptual dominion over earlier ones, they may not entirely replace them. They may, in fact, coexist and blend despite the sometimes obvious conceptual incoherence. Then again, a blending of compatible metaphors may enhance understanding through elevated coherence. Each metaphor is accompanied by a specific narrative allowing the metaphor to undergo explanatory elaboration and expanded relevance. Personal metaphors also have significant organizing power as self-referencing and self-regulating metaphors. New conceptualizations suggest two additional root metaphors: First, *biotopism* (Greek, bios = life + Greek, topos = place), the view that human development is governed by non-linear dynamic systems (van Geert, 1993, p. 265); and second the *physical point of view*, the belief, informed by neuropsychology, that the human mental apparatus can be described from the physical point of view in which brain functionality is integrated with core psychoanalytic concepts (Kaplan-Solms and Solms, 2002BIB-206, p. 251).

5 Strenger (1997) offers "critical pluralism" as a solution to the tension between purism and pragmatism. "Purist approaches have a unitary clinical style that is integrated with a theory of human nature and development. Pragmatism is the belief that no map matches the territory, and that it is dangerous to sacrifice the patient's interests on the altar of unified conceptions" (p. 111). Purism is the dominant tradition in psychoanalysis. Arguing for a complexity of perspectives, the task of the clinician is to minimize intuition and gut reactions in the face of helplessness. The "critical" dimension in critical pluralism refers to maintaining a creative inner space that can contain a multiplicity of perspectives.

5
CASE FORMULATION FROM A REGULATION THEORY PERSPECTIVE

Regulation theory can be operationalized by using a construct incorporating specific elements in a flexible overarching theoretical system. This operational construct might be described metaphorically as constituting a developmental algorithm (Bendicsen, 2013, p. 196). This developmental algorithm consists of seven overlapping and complementary domains of knowledge: 1) modern metaphor theory; 2) attachment theory; 3) self psychology with intersubjectivity theory and relational psychoanalysis; 4) cognition; 5) contemporary psychoanalytic developmental psychology; 6) complexity theory; and 7) neurobiology with narrative theory. I will construct a case formulation using selected dimensions from these domains to form a comprehensive explanatory hypothesis. In order to adequately explain other cases, different configurations of other domains of knowledge might be required.

Case formulation using a developmental algorithm

Modern metaphor theory

The significance of metaphor in human experience cannot be overstated. "The drive toward the formation of metaphors is the fundamental human drive, which one cannot for a single instant dispense with in thought, for one would thereby dispense with man himself" (Nietzsche, 1873/2010, p. 42, paraphrased). The spontaneous emergence of the self-referencing metaphor, "Top dog on the floor," is understood as a most significant event in both development and treatment. Unlike classical metaphor theory which locates metaphor in words and language, modern metaphor theory, due directly to neurobiological research findings, locates metaphor in body functionality and body tissues. So metaphor now is thought of as an embodied process which can lend personality tonus (Blos, 1962, p. 129; Panksepp,

1998, p. 149; Lakoff and Johnson, 1980/2003, pp. 255–256) to the new set of potentials and possibilities awakened in the late adolescent. Whereas Blos was referring to an optimal degree of anxiety favoring developmental consolidation or a necessary state of tension that accompanies the stable personality, metaphoric personality tonus is used here to refer to the concept of transformational readiness in the transition from late adolescence into young adulthood. Transformational readiness of the self-state occurs when the certitude of self-doubt and inaction is replaced by the ambiguity of possibility with associated action potentials (Bendicsen, 1992). Tonus or tone, comes from the Latin for tension, and is defined as, "The partial steady contraction of muscle that determines tonicity or firmness; the opposite of clonus" (Taber's, 1997, p. 1968). In this regard, Lichtenberg, Lachmann and Fosshage (2011) refer to states of "dialectic tension" amongst motivational systems that influence behaviors shaping aspirations and goals.

Embodiment of metaphor generates a natural linkage with motivation. Panksepp, writing about how the activation of his SEEKING system modifies subjective experience, especially intense interest, engaged curiosity and eager anticipation, began to draw a distinction between personality traits as opposed to emotional states. Granting that interest, curiosity and anticipation may not fit the precise definition of an "emotion," the psychobehavioral qualities associated with these experiences are pertinent. Suggesting that curiosity and interest seem to be relatively stable personality traits or tendencies as opposed to passing emotional states, Panksepp mentioned, "In fact, contrary to most other emotional responses, the SEEKING system is commonly tonically engaged rather than phasically active" (Panksepp, 1998, p. 149).

Modell's (1997 and 2000 in Bendicsen, 2013, pp. 81–82) concept of the bifurcation of metaphors into open and foreclosed varieties is relevant here. Open metaphor refers to the individual's capacity for recontextualization of affect and reconfiguration of meaning in the reevaluation of experience. In foreclosed metaphor categories of affect are remembered and repeated, not recontextualized. There may be "a telescoping of time so that the affective experience of past and present are identical" (Bendicsen, 2013, p. 81). The appearance in Myles' treatment of a self-referencing metaphor of the open variety, in the context of a secure attachment experience, heralds a shift in thematic content and suggests that an advance in growth/progress may be expected.

I contend that the spontaneous appearance of a self-referencing metaphor can provide, for the late adolescent, a shape and direction to new identifications thus guiding development. Hustvedt (2016, p. 361) captures this teleological sense perfectly when she quotes Lakoff and Johnson on the nature of modern metaphor:

> Metaphors may create realities for us, especially social realities. A metaphor may thus be a guide for future action. Such actions will, of course, fit the metaphor. This will, in turn, reinforce the power of the metaphor to make experience coherent. In this sense metaphors can be self-fulfilling prophesies.
> (Lakoff and Johnson, 1980, p. 1156; see also Levin, 2009)

In addition to the embodiment of metaphor and motivation, Merleau-Ponty, the French philosopher, believed that the body was the primary site of the source of knowledge. Replacing consciousness as the source of knowledge, he maintained that the body and that which it perceived could not be disentangled from each other. Under Merleau-Ponty, intersubjectivity and phenomenology were experiences best understood from an embodied perspective.

The profound significance of metaphor in our lives takes on greater relevance when in child developmental research on cross-modal sensory perception we learn that we are born without sensory boundaries. Distinct sensory perceptions for the infant begin as a blur and then gradually differentiate, accelerating with language.

> This would help explain the many forms of synesthesia people experience, hearing colors, seeing letters and numbers as colors, or feeling sounds. Synesthetes retain cross-modal experience that other people lose. But to one degree or another intermodal perception is part of all our lives. Metaphor jumps across the senses all the time. I am feeling blue. Listen to that thin sweet sound. What a sad color.
>
> (Hustvedt, 2016, p. 244)

To reaffirm the significance of the self-referencing metaphor in development, in another example, we need only to remember that the seventeen year old Freud, as he was beginning his medical education, had developed a strong admiration for and identification with Darwin. From birth Freud was marked by his parents as destined for greatness. Freud identified with a variety of heroes as his development unfolded: Hannibal, Cromwell, Napoleon, Darwin, Cortez and Schliemann (Sulloway, 1979, pp. 476–480; Gay, 1988, pp. 172, 326). Schimmel (2014) maintains that by age 43, Freud had developed dual self-referencing metaphors that guided him to the highpoint of his career in 1900 with the publication of the *Interpretation of Dreams*. The metaphors he used were that of the theoretical conquistador and the thinker, the intellectual, daring discoverer and the conceptual synthesizer. At about the halfway mark in Freud's life an integration and stabilization of the self had occurred to include the adhesive identifications of conqueror in balance with the psycho-synthesizer, twin identifications which were to guide him the rest of his life.

Attachment theory

The traditional idea that the end of adolescence should usher in a state of autonomy is no longer accepted. The separation-individuation (Mahler, Pine and Bergman, 1975) and the second individuation (Blos, 1967) concepts with their implication of independence from oedipal (psychologically incestuous) family relationships as a developmental, culturally driven goal, is understood now as overemphasized. A more appropriate and realistic appraisal of end of adolescent phase relationships is captured in the phrase attachment-individuation (Lyons-Ruth, 1991). Referring to Mahler's infant research, Lyons-Ruth states,

> The central developmental thrust that Mahler was attempting to capture might be better thought of as an attachment-individuation process rather than a separation-individuation process. The attachment individuation terminology emphasizes the infant's propensity to establish and preserve emotional ties to preferred caregivers at all costs, while simultaneously attempting to find a place within these relationships for his or her own goals and initiatives (see Lichtenberg, 1989). This emphasis on the overriding nature of the infant and toddler's goal of maintaining access to a preferred partner or partners at times of stress is not clearly indicated within the separation-individuation terminology.
>
> (Lyons-Ruth, 1991, p. 10)

Doctors (2000) applies attachment-individuation to the adolescent phase.

> I expand on Lyons-Ruth (1991) and Doctors' (2000) felicitous phrase "attachment-individuation" to convey the idea that positive intersubjective connectedness to phase-specific selfobjects is a developmental imperative if the person (i.e., the adolescent) is to attain and maintain a functional, adaptive sense of competence in her place in society.
>
> (Bendicsen, 2013, p. 53)

It is recognized that late adolescents and young adults thrive when embedded in family and peer relationships, those that are recalibrated to enhance individuation strivings. Others, as selfobjects, are needed throughout life. In addition to the self as being acted upon, getting needs met by the object as a selfobject, "there is another group of phenomenon that proceed in the opposite direction, that is, phenomenon characterized by the self as actor and the self-object as the acted upon" (Wolf, 1988, p. 60 in Bendicsen 2013, pp. 190–191).

But can we be more precise about what is meant by individuation? Palombo (2016), in writing about the self as a complex adaptive system, offers a distinction between differentiation and individuation that is helpful.

> The vigor and vitality that a child manifests during development reflects the maturity the child attains. Differentiation involves, among other experiences, the joyous encounters with life's challenges, the consolidation of the self in adolescence, the ambitious pursuit of an ideal in adulthood, the creative generativity of maturity, the sense of being an independent center of initiative, and the possessor of a coherent narrative.
>
> (Palombo, 2016, pp. 12–13)

In formulating this position Palombo seems to have been influenced by Kohut's notions of the transformations of narcissism, positive mental health and normality (Kohut 1966/1978; 1977; and 1984; and Palombo, Bendicsen and Koch, 2009, pp. 257–281).

> Differentiation occurs at multiple levels. It is observable at the neurophysiological level in the brain changes that occur through maturation, at the psychological level in the complexity of our capacity to process thoughts and feelings, and at the social level in the entangled relationships that we develop with others.
>
> (Palombo, 2016, p. 13)

In addition:

> The sense of agency is associated with the capacity to effect changes in ourselves and in the context that we inhabit. We associate it with the experience of being effective as agents of change, and the experience that Basch (1980) called the feeling of competence.
>
> (p. 13)

With respect to individuation, Palombo (2016) offers, "The uniqueness of each individual challenges us to explain how it is that each person retains his or her individuality even as each is in a continuous state of flux" (p. 14). The sense of self is in a continuous state of reconfiguration, organized around attractor states. "The central enigma we face is how to delineate our individuality without particularizing our separateness" (p. 14). As our sense of self changes, so too are the multiple contexts in which we exist. "As social beings, isolation from others deprives us of the psychological nourishment that we need in order to survive" (p. 15). Palombo sets "attachment" in a non-linear dynamic systems perspective and now prefers to use the term "interconnectedness" (Personal communication, December 11, 2016).

So I take it that in Palombo's distinction, differentiation refers more to the intrapsychic matrix while individuation pertains more to sociological experience, both with dynamics contextualized and understood in terms of complex adaptive systems. With both realms being vital and necessary it then seems useful to refer to and otherwise characterize the late adolescent sense of self in a developmental condition of attachment-differentiation/individuation. Emphasizing the developmental needs for both connection and autonomy in the emerging adult Viner has coined the expression "connected autonomy" (Monroe-Cook, 2016, p. 18).

Myles, having experienced a secure attachment, is sustained by his deep involvement in a mutually supportive network of functional family relationships (mutually reciprocating selfobjects), those that are also there for him in crisis. Myles' uncle, age 27, is challenged with some questions about next step issues in his career and love relationship. He has used Myles as a selfobject to reflect on this situation, help sooth his anxious state and formulate a plan of action. This efficacy experience (Wolf, 1988, p. 60), where the self experiences selfobject success as an agent in influencing the object, is understood as a narcissistic enhancement. Mature differentiation/individuation of self from objects is heightened. Myles' self-worth is validated and his stance in the world as a useful, caring individual is reaffirmed.

Self psychology with intersubjectivity theory and relational psychoanalysis

Kohut (1959; 1966; 1971; 1977) and his core collaborators succeeded in creating an alternative theoretical framework that overthrew the dominance of ego psychology. Positing a separate line of development for narcissism per se, narcissism is now considered normal with healthy end points: creativity, humor, empathy, self-calming and regulatory capacity, wisdom and an acceptance of the finality of life. Empathy, defined as vicarious introspection, is now regarded, among other things, as a research instrument. Significant concepts include recognition of three sets of developmental needs: affirming and admiring; safety, regulation and soothing; and commonality. Each set of needs dovetails with different transferences respectively: mirroring, idealizing and alter-ego or twinning, corresponding to the constituents of the tripolar self. The selfobject formulation references caregivers in the child's life who provide life-sustaining functions which gradually become transmuted into psychological structure. Selfobjects are never outgrown, but optimally change in accord with developmental needs (Palombo, Bendicsen and Koch, 2009, pp. 257–281). The self-selfobject matrix relationship is now understood to cradle the subjective world through the embodied processing, categorizing and meaning-making of experience. Wolf (1988) makes the point that the self-selfobject relationship proceeds in both directions with functions provided to the self by the selfobject and, alternately, the selfobject derives benefit from functions provided by the self. In addition, the self of self psychology breathes metaphoric life into the neural self.

Myles has demonstrated a clear ability to use reciprocating selfobject experiences to maintain vigor, cohesion and adaptability/resilience. Myles' "condition," as he has labeled it, is managed through the continuous construction and reconstruction or stabilization and restabilization of his self-states as his "condition" fluctuates. When his "condition" fluctuates and intensifies, the threat of fragmentation of the self leads to the desperate search for dependable selfobjects. This search is experienced as a counterweight to possible behavioral enfeeblement and terrifying disintegration anxiety. Modulation and regulation of this anxiety becomes the essential therapeutic task.

Let us reference Palombo's outline of the development of the self. During late adolescence (18–20) a number of shifts in the adolescent's sense of self are observed (Palombo, Bendicsen and Koch, 2009, pp. 274–276). I will list these and then relate some in particular with Myles' developmental progress.

a) Painful self-consciousness begins to dissipate.
b) Egocentrism gives way to more empathic attitudes towards others.
c) Self-regulation becomes internalized.
d) Affect states become less labile.
e) Greater self-confidence is manifested.
f) Self-assertion without hostility is more observable.
g) Regressions are less frequent and less severe.

h) Experimentation with fringe activities due to peer pressure lessens.
i) Fantasy is used more for creativity or trial action than for defensive purposes.

Comparing Myles' shifts in his sense of self to the shifts in the generic adolescent's sense of self easily suggests that Myles, at age 21, is a mature young adult. However, two shifts stand out as relevant to Myles' "condition." First, in the face of psychotic process the lack of internalized self-regulation becomes both a feature of the brain disease and, its strengthening, a goal of treatment.

> Whereas normal states of awareness are comprised of an integration and a balance of left and right hemisphere processing, psychosis may be a result of the intrusion of right hemisphere functioning into conscious awareness. Hyperactivation of the right hemisphere, or a decrease in the inhibitory capacities of the left, may diminish the ability to filter primary process input from the right hemisphere. This shift in right-left bias may occur for many reasons, including changes in levels of important neurochemicals such as dopamine, neurochemical abnormalities, or changing activation in subcortical brain areas such as the thalamus. Schizophrenic patients and their close relatives demonstrate reduced left hemisphere volumes in the hippocampus and the amygdala, which has shown to correlate with thought disorder. (Seidman et al., 1999, Shenton et al., 1992).
>
> (Cozolino, 2010, p. 108)

The right-left hemispheric filtering process hypothesis was earlier echoed by Claridge, Pryor and Watkins (1998). Attempting to understand the psychological functioning of the schizophrenic brain, they state that the two hemispheres are engaged in a continuous, cooperative exercise with each contributing to specialized tasks. With the left dominant in speech and language, each half has adopted different modes of information processing.

> The left, consistent with its primary language capacity, is sequential and analytic in its style and has often been regarded as the rational half, responsible for the organization and expression of conscious thought. The right hemisphere has a more global, less focused, free ranging style and, being "silent," can only influence conscious awareness indirectly.

Corresponding to the two processing modes, two stages of information analysis are suggested.

> The right hemisphere is well-placed to undertake the early parallel, preconscious scanning of large amounts of information, the left taking on the later function of conscious elaboration of selected items. The delicate balance of cooperation between them – occurring across the *corpus callosum* and involving some kind of filtering process – is therefore of crucial importance in allowing

the left hemisphere to remain responsive to the right hemisphere's influence, yet sufficiently in control to avoid overloading of conscious, direct thought. The authors are in agreement with *psychological* research that the unusual information processing found in psychotic individuals is an exact parallel in theories emphasizing the functioning of the corpus callosum as the *physiological* vehicle for psychosis.

(pp. 43–45)

In studying the right brain-left brain relationship from a pathological perspective it is helpful to keep in mind the normal right brain-left brain relationship. Schore (2011) writes that the idea of a single, unitary brain is misleading.

> The left and right hemispheres process information in their own unique fashion and represent a conscious left brain self system and an unconscious right brain self system. Despite the designation of the verbal left hemisphere as "dominant" due to its capacities for explicitly processing language functions, it is the emotion processing right hemisphere and its implicit homeostatic-survival and communication functions that is truly dominant in human existence. (Schore, 2003b). The early–forming implicit self continues to develop, and it operates in qualitatively different ways from later-forming conscious explicit self.
>
> (Schore, 2011, pp. 76–77)

Regarding:

> Auditory hallucinations, or hearing one or more voices talking, are a core symptom of schizophrenia ... These aberrant, intrusive, and ego-dystonic experiences may reflect right hemisphere language (related to primary process thinking and/or implicit memories) breaking into left hemisphere awareness. These voices, often heard as single words with strong emotional value, are experienced as coming from outside the self.
>
> (Cozolino, 2010, p. 108)

Second, with respect to Myles' "condition," the subject of regressions is directly related to his brain disease. There appears to be a direct functional relationship between the fluctuation in his condition and his behavioral regressions. As the brain disease intensifies, he becomes vulnerable to regressive phenomenon.

In the young adult (21–26) the central shifts in the sense of self involve:

a) functioning more as an independent source of initiative;
b) arriving at an alignment among one's goals, capacity to achieve those goals and the and motivation to progress toward completion; and
c) developing the resilience in self-esteem to counter the impediments toward achievement. (Palombo, Bendicsen and Koch, 1991, pp. 274–276)

Because from a chronological point of view, Myles is on the border between late adolescence and young adulthood, we need to consider both sets of shifts in the self. In Myles' situation, the employment and relationship data suggest that he is functioning as a mature young adult. However, the brain disease is the monolithic factor compromising his life and dominating his existence. Of course, beneath his mature functioning lies a profound and brittle vulnerability. The massive trauma of the schizophrenic break and its terrifying sequelae brought him to the edge of annihilation of the self. During weakened moments reliving the trauma must evoke the defense of dissociation as a way to compartmentalize self-states that defy integration. Such data, however, has not made its way into our therapeutic process. What keeps the trauma from receding into a self-state of affective numbing chaos? Bromberg (2003) offers an explanation. The key to keeping his selfhood safe and stable is to increase self-reflective capacity and keep in consciousness a dimension of self-reference so that his "condition" continues to belong to him. It remains symbolized as an endogenous dark force that is a "basic fault" in his constitution that requires constant vigilance and self-care. Kept in conscious awareness, "a link must be must be made between the mental representation of the event and a mental representation of the *self* as the agent or experiencer" (Bromberg, 2003, p. 563). If I understand Bromberg correctly, he is saying that if the mental representation of the event resides in working memory and is not split off or dissociated, it is available to Myles as mental content that can be re-experienced and processed. He can then proceed to deliberately respond to his needs by immersing himself in counterbalancing, mutual, co-regulating therapeutic activities. Myles draws on his self-referencing metaphor, "Top dog on the floor," to find the strength to adapt with resilience to life's daily demands to stay regulated in a healthy selfobject milieu. *His self-referencing metaphor consolidates personal identity by strengthening his autobiographical self.*

The neurobiological research suggests that the different self systems are kept in balance, in normal functioning individuals, by a filtering mechanism, quite likely located in the corpus callosom, that links the two hemispheres in a *regulatory* process that allows the left hemisphere to suppress primary process intrusions from the right hemisphere.

Supportive relationships

The question of how supportive people in our lives help us through rough patches merits our attention, especially in view of contemporary perspectives. In psychoanalytic theorizing how have supporting relationships been characterized? Among the earliest theoretical frameworks to elevate everyday sustaining relationships alongside conventional office-bound therapeutic efforts in clinical work was ego psychology. Ego psychology understood ego functions (Brenner, 1973, pp. 31–126), from a developmental perspective, as evolving through gradual differentiation from the id, operating largely out of awareness, to become internalized. Ego defenses are an ego function. Ego functions are pathways through which the individual adapts to the external world: "the ego not only has been described as a group of functions, but also as both a complex self-regulating structural organization and a motivational system" (Goldstein, 1984, p. 53). Goldstein (pp. 53–71) has a more thorough outline of these ego functions and draws a useful distinction between ego-supportive (improving the individual's adaptive functioning) and ego-modifying (changing basic personality patterns or structure) interventions (p. 166). Supportive others provide ego assistance and when in alignment with the aims of the ego are called ego allies.

Turning to self psychology, ego-oriented language has been replaced by an adaptive system in which multiple, successive selfobjects accompany the self through development. How is the selfobject currently conceptualized, recognizing that it is not uncommon for popular terminology in psychoanalysis, with increased usage over time, to undergo conceptual broadening? Schechter (2014, pp. 174–175) reminds us that Kohut intended the selfobject to be understood as the experience of an unconscious, internalized function or set of functions, "a purely 'intrapsychic' stance." However, some of his followers struggled with the external "object" in selfobject and came to the opinion that Kohut lost sight of the importance of the actual other who provides the psychological assistance. I take the position that psychological nutrients can be metabolized into cohesive self-structure through both the interpretation of empathic failures in the transference *and* through accentuating the healing qualities in the relationship to be a new and better object for the patient. Terman put it well: "it is the repetition of the presence [of help] that builds new structure. The repetition of the absence re-evokes the old patterns" (Terman, 1988, p. 121 in Schechter, 2014, p. 175).

Newman (1992) contributed significantly to the definitional broadening of supportive people with his immanently useful term, the "usable object." With conceptual support primarily from Winnicott (1971), Bacal (1985), Kohut (1977, 1984), Terman (1988) and Hoffman (1983), Newman moved away from the classical psychoanalytic understandings associated with one-person psychology and embraced a two-person psychology stance in which intersubjectivity informs transference as a two-party system. The term "usable," "first referred to by Winnicott (1971), implies an object whom the patient views as separate and as a source of emotional sustenance for the matrix of the self" (Newman, 1992, p. 132). Newman is addressing a particular demographic: "Most of our patients live in an internal atmosphere chronically peopled by malignant introjects or selfobjects" (p. 131). In this two-person psychology, Newman calls for selected gratification of legitimate patient needs in a new context of being an authentic object in the real relationship. The "difficult" patient experiences the therapist as a mutative new object in an empathic milieu of optimal responsiveness, replacing optimal frustration as the operative therapeutic stance. In Winnicott's view "usability" refers to a new or corrective developmental experience, a positive connection, in which the patient is provided with an environment of usability in the forging of a new object experience (p. 132).

In more of a generic stance, Galatzer-Levy and Cohler (1993) define the "essential other" from a postmodern, social constructivism position. The essential other

> is our experience of other people, and entities in the environment, that supports the sense of a coherent and vigorous self and its development. The essential other refers to the experience of the psychological life of the individual, not to mention the external reality of these people. We believe that the support of the self is always part of a total experience of other people and entities. This function is never isolated from the additional meaning these others have.
>
> (Galatzer-Levy and Cohler, 1993, p. 3)

For our purposes, Galatzer-Levy and Cohler emphasize the dual nature of supportive others including: 1) the reality of concrete assistance they offer and 2) the subjective experience of dimensions of supportive others in states of co-regulation that have become internalized.

So whether one talks about ego allies, selfobjects, usable objects or essential others, these concepts attempt to capture the multiple features of and dynamics associated with sustaining the self through normal development as well as in formal clinical circumstances. In the case of Myles, whether my therapeutic role is that of an ego ally or a calming selfobject, the repetition of the real relationship in the present is, without question, the critical factor in building and maintaining his self-regulatory capacities.

Cognitive theory

The cognition traditionally associated with late adolescence has expanded beyond Piaget's Formal Operations to a variety of Post Formal Operational thought models. Piaget's step-wise cognitive model is usually understood in the view from a far perspective where progress of cohorts is grouped and averaged. Alternatively, the view from below privileges the individual model and has more relevance to the single case study and research methodology. This dual perspective arises out of non-linear dynamic systems theory or complexity theory, a nontraditional way to approach cognitive developmental model thinking. A complexity theory framework is useful in this case study in that Myles presents with a unique biopsychosocial profile organized predominantly around biological brain neuropathology. In complexity theory human motivation may not be psychologically oriented due to strange attractors, system organizers, exerting unknown or unpredictable perturbations on the system. "Consideration of this possibility spares us from looking for motives for shifts when intrinsic characteristics leads to those shifts" (Galatzer-Levy, 2004, pp. 428–429). Stage theory is replaced by states and processes (Thelen and Smith, 1994 and Bendicsen, 2013, pp. 133–147, 188–189). When Myles' "condition" is stable and non-refractory, it is clear he has above-average intelligence that he can harness in the service of optimal functioning in the competitive world of retail sales. Because this skill set is organized around verbal relationship interaction, he can experience success. However, his visual disturbance is so significant that it compromises traditional learning which relies on reading comprehension. Therefore, multiple self-state attractors or organizers compete for dominance.

A second concept, that of hot (active) and cold (passive) cognitions, merits our attention. "Hot cognitions are emotionally biased brain activity that can override rational thinking, leading to risky behavior" (Spear, 2010, in Bendicsen, 2013, p. 189). Hot cognitions are heat-of-the-moment decisions emerging from increased activity in the amygdala and other subcortical regions necessary for rapid and instinctual behavioral responses. While not restricted to adolescence, hot cognitions are prevalent during this developmental phase. Cold cognitions, on the other hand, are associated with attenuating activity in the prefrontal cortex regions critical for logical thinking and executive control (Arnsten, 1998 in Spear, 2010, pp. 187–188). Cold cognitions are related to top-down brain function modeling while hot cognitions are those connected to developmentally immature or bottom-up functionality. The appearance of a new affectively charged self-referencing metaphor provides a synergistic pull to move forward in development, in this case, into young adulthood. Myles has acquired such a cold cognition in the form of "Top dog on the floor" self-referencing metaphor. This appellation assumes risk taking in the form of an adjustment to a new identification; it organizes him and offers focus to the developmental move into the next era of life.

A third dimension, that of cognitive dissonance, a type of psychic conflict, is of value here in that as new self-referencing schemas arrive, they can either blend or clash with the old. In Myles' case, old schemas of "I am defective," "I am going nowhere" and "Nobody cares about me" clash with the grandiose and narcissistic designation, "Top dog on the floor." The mind in search of health and harmony will not tolerate such cognitive dissonance and so must struggle to integrate the new schemas into a realistic agenda for growth. In Myles' situation this process is proceeding and moving forward. It is as if Myles is trying to encapsulate the "condition" and so move ahead with those aspects of his life that he can direct. This type of cognitive dissonance is associated with hot and cold cognitive processes and so is more maturational and biological in nature.

The second type of cognitive dissonance is distinctly psychological in nature and is aligned with the consolidation of different selves into a coherent and cohesive self-state that marks the delineation of the young adult character structure. Lieberman (2007) has summarized this position.

> Cognitive dissonance research has established that when individuals perform a behavior or make a choice that conflicts with a previously established attitude, the attitude tends to change in the direction that resolves the conflict with the behavior. From the outside the process appears to involve rationalization, whereby individuals strategically change their attitudes to avoid appearing inconsistent.
>
> (p. 273)

Additionally, this process seems to involve contributions from conscious and unconscious dimensions.

A fourth concept concerns decision making under emotional strain. Recent cognitive science research investigated the neural basis of one way that affect modulates cognition. Specifically, the research suggests that positive mood enhances both insight (sudden, creative cognition) and analytic (gradual application of deliberate solution strategies) problem solving ... by modulating attentional and cognitive control mechanisms within the anterior cingulate complex. While positive mood enhances both problem-solving methods, a strong bias has been found suggesting a positive association between insight problem solving and positive mood. Insight problem solving appears to be associated with stronger contributions from the right hemisphere, while analytic problem solving is linked to greater activity in the left hemisphere. As expected, negative mood, including depression and anxiety, is associated with deficits in attentional and control mechanisms (Subramanian et al., 2009). Another study by Heilman et al. (2010) examined two responses to decision making under risk and uncertainty. They were *cognitive reappraisal*, an antecedent-focused emotion regulation strategy that alters the trajectory of emotional responses by reformulating the meaning of the situation and *expressive suppression*, a response-focused strategy that involves inhibiting behaviors associated with emotional responding. The data suggested that reappraisal is effective in reducing the experience of negative emotions while suppression of negative emotions accompanied by high arousal is associated with impulsive decision making (pp. 257–258). Considering the chronic emotional strain Myles is under it seems reasonable to assume that he has found ways to enhance positive mood through cognitive problem-solving strategies that provide him a sense of efficacy in mastering stressful work circumstances. The thoughtful, anticipatory method of decision making he employs suggests a reliance on cognitive reappraisal to maneuver through the complexities of remaining successful in his competitive, often exhausting environment.

As a part of the discussion on emotional decision making when under emotional strain, it should be mentioned that research on the neurocognitive basis of schizophrenia as an information-processing abnormality is proceeding (Aleman, 2014). Aleman summarizes the research in a meta-analysis on cognitive impairment in schizophrenia: "Impaired activation and connectivity between frontotemporal, frontoparietal, and frontostriatal brain networks subserving cognitive functioning and integration of cognition and emotions has been consistently reported" (p. 115). In the *Diagnostic and Statistical Manual of Mental Disorders*, DSM-5 (2013), cognitive impairment is categorized as "Disorganized Thinking" and is implied from the individual's speech. Disturbances in thought patterns are inferred from the degree of coherence/incoherence in speech. Significant cognitive deficits in attention, memory, executive functioning and general intellectual abilities are widely noted in individuals with schizophrenia (p. 88). In Myles' case cognitive functioning, inferred from clarity of speech, has been remarkably coherent. However, into the fifty-fourth month of treatment (4.5 years into treatment) cognitive slippage is noted with criticism from his supervisor in his lack of attention to core details in his primary assignment and that he has unilaterally grown expansive in taking on unnecessary responsibilities in other departments. This is consistent with manic episode involvement. Close engagement with trusted selfobjects in his chain of command will be vital in mitigating this trend.

Contemporary psychoanalytic developmental psychology

The developmental model

Interest is accelerating on crafting a contemporary psychoanalytic developmental model. Progress in a variety of knowledge domains has generated a push for integration of these domain variables. If an integration is too ambitious, might a call for an alignment of such variables be more realistic, such that they contribute toward a coherent roadmap offering an explanation of the role each plays in our current understanding? The first psychoanalytic developmental model, Freud's monumental psychosexual synthesis (Freud, 1905), was fashioned in the context of a positivist philosophical orientation with stages unfolding in a progression that is linear, invariant and sequential. Even reading this framework today leaves the impression of exactitude, completeness and certitude (Palombo, Bendicsen and Koch, 2009, pp. 24–33). In today's postmodern philosophical orientation, subjectivity and understanding/interpretation replaces objectivity and predictability (Bendicsen, 2013, pp. 203–207). The place of developmental theory in psychoanalysis ranks second only to clinical experience in shaping analytic theory and practice (Galatzer-Levy, 2017, p. 64). Change is enveloping developmental conceptualization today where extensive debate is occurring in consideration of a positivist perspective versus one informed by postmodern thinking and relational theory viewpoints (Tyson, 2002; Rangell, 2002; Gilmore, 2008, Dick and Muller, 2018).

To advance this discussion let us consider the likely elements in a contemporary developmental theory that Gilmore suggests.

A psychoanalytic developmental theory (Gilmore, 2008, pp. 897–900) should:

1) "maintain its focus on the body and bodily impulses and the powerful impact of the maturing body on the mental life of the individual."
2) with respect to a bodily focus, "appreciate the impact of the developing gendered body on the mind."
3) examine the ongoing interaction of endowment, environment and experience that reverberates throughout the transformations in mental organization occurring in different developmental epochs, including cognitive capacities and defense mechanisms.
4) understand the development of the sense of self, self-regulation and subjective experience of identity for all stages of human life and for all theoretical persuasions.
5) recognize the demands imposed by the near and larger cultural environments, including cognitive and social expectations, of each developmental epoch.
6) recognize the contribution that adaptation in the family system plays in the individual's unfolding life.
7) maintain its focus on the dynamic unconscious, sexual and aggressive impulses, object relations, ego capacities including defenses, self-experience and identity, and explicate the way mental structures are built up across developmental epochs.

8) maintain an open posture toward information obtained from neighboring developmental sciences.

Paraphrasing Gilmore, the eight elements in a contemporary psychoanalytic developmental model should include: 1) embodied psychological/biological processes, 2) gender influence on mind, 3) the influence of endowment, environment and experience throughout the life cycle, 4) development of self, self-regulation and subjectivity understood from a multiplicity of theoretical orientations, 5) the impact of culture, 6) the adaptation to the family system, 7) the centrality of the dynamic unconscious, sexual and aggressive impulses, object relations, ego capacities including defenses, self-experience and identity, and explicate the way mental structures are built up through the life cycle and 8) receptivity to nonpsychoanalytic interdisciplinary approaches. With respect to the last element, in order to have broad appeal, three elements need to be added: a contemporary theory of human motivation, neurobiological research findings and non-linear dynamic systems theory.

Let us look at this subject from Tyson's (2002) perspective, who in a scholarly article, addressed the difficult task of integrating long-held psychoanalytic principles with nonpsychoanalytic domains of knowledge. As these domains are examined, it becomes obvious that scientific empirical data fails to support many treasured psychoanalytic concepts described in metaphorical, philosophical terminology. If the dynamic unconscious and conscious processes are characterized by primary and secondary process thinking, how is it possible to integrate the neurobiological unconscious and conscious processes characterized by implicit memory (procedural, emotional, nonverbal, not symbolically encoded memories, more right brain) and explicit memory (autobiographical, verbal, symbolic memories, more left brain) (p. 28)? If one attempts to integrate these two positions, what would the conceptual bridges look like?

Let us take another example. Freud conceptualized human motivation as explained by psychic energy, the dual-drive theory and the structural theory in which anxiety eventually was understood as an ego affect (Tyson, 2002, p. 24). Contemporary thinking about motivation now must include, among others, these dimensions: attachment theory, non-linear dynamic systems theory, affect regulation, the self as agent, self-regulation, understanding the self as a complex adaptive system (Palombo, 2013a; 2013b; 2016; 2017a) and a life cycle framework.

Tyson includes in her developmental theory the centrality of the role of affect as a "signal function, i.e., affect recognition, and self-reflection in affect regulation" (Tyson, 2002, p. 41). She concludes that from a cross-disciplinary perspective a revised theory of affect is within reach. She also sees a place for neurobiology and non-linear dynamic systems theory, with non-linear dynamic systems theory posing the most difficult integrative challenge. So the pieces are gathering, but how would they fit into an overall coherent framework?

As we think about this issue, I am reminded of Tronick's (2007, pp. 4–5) cautionary emphasis on the need for clarity of definitions when applying concepts across disciplines: "Conceptual mixing can lead to eclecticism where the outcome can be experience distant, such as in neuroscience models of the brain where the individual is absent."

Considering the enormous complexity and difficulty integrating psychoanalytic and nonpsychoanalytic perspectives in just these two examples, I advocate bypassing energy-consuming attempts at integration, opting instead for a pluralistic approach (as outlined in Chapter Four). As mentioned, a pluralistic approach eschews the struggle to blend terminology from incompatible disciplines, bypasses attempts to integrate vastly disparate data from different knowledge sources and sidesteps issues of priority and the scaffolding of theories. It allows diversity of perspectives and emphasizes flexible, contextual alignment of theories that offers the optimal explanatory power at a point in time.

It seems to me that Palombo's four-part work (2013a; 2013b; 2016; 2017a), explicating the self as a complex adaptive system, offers us an opportunity to compare and contrast some differences between pluralism and scientific realism. Let us examine his contribution in some detail. Palombo starts with the overall theme of constructing an interdisciplinary contemporary model of human development that is grounded in data-informed scientific realism and is psychoanalytically oriented. He declares that "A proposed framework must integrate biological/neuropsychological factors, introspective factors that deal with the meanings the patient construes from his or her own experiences (Sander, 1995), and interpersonal (social-emotional/relational) factors related to the context in which patients live" (Palombo, 2012, p. 4). This ambitious undertaking is to be achieved through an "integration" of these factors. While "integration" is not defined, I infer that an "alignment" of theoretical variables is not excluded from the "integration." The absence of a definition of "integration," it seems to me blurs the distinction between scientific realism and "scientific pluralism" (Goldberg, 2007, p. 1673). While Palombo may be arguing for a goal of an eventual seamless union or unity of knowledge domains, he may also be enlarging the theoretical tent in an effort to build a comprehensive model.

Before we examine Palombo's construct it is helpful to clarify what is meant by scientific realism and scientific pluralism. Historically, scientific realism has been associated with movements to unify knowledge as in monism, "the view that reality is one unitary organic whole with no independent parts" (Webster's New Collegiate Dictionary, 1981, p. 737), to engage in the positivist search for truth or the approximate truth of scientific theories, and to produce true descriptions of things in the world. "Scientific realism is a positive epistemic attitude toward the content of our best theories and models, recommending belief in both observable and unobservable aspects of the world described by the sciences" (Stanford, 2017, p. 1). There are three dimensions of scientific realism. "Metaphysically, realism is committed to the mind-independent existence of the world investigated by the sciences" (p. 2). "Semantically, realism is committed to a literal interpretation of scientific claims about the world. In common parlance, realists take theoretical statements at 'face value'" (p. 3). "Epistemologically, realism is committed to the idea that theoretical claims (interpreted literally as describing a mind-independent reality) constitute knowledge of the world" (p. 3).

On the other hand, scientific pluralism is often considered to be antirealism because of its embrace of multiple descriptions of things in the world reflecting the complexity of the subject matter. Dickson (2006) has defined scientific pluralism as "the existence or toleration of a diversity of theories, interpretations, or methodologies within science" (p. 42). Pluralism can be divided into competitive and compatible dimensions. The competitive dimension has been dominant and seen as desirable because it is considered necessary in order to hasten progress in science. Scientific growth occurs as a result of exposing a set of research hypotheses to rigorous empirical scrutiny and analysis thereby increasing the probability that epistemological fallibility will be minimized. It is considered "the rational strategy to adopt for the scientific community as a whole in order to hedge its bets against empirical uncertainty" (Mitchell, 2002, p. 56). In endorsing competitive pluralism, "the ultimate aim of science is to resolve the conflicts by adopting the single unfalsified, or true, or overwhelmingly supported winner of the competition" (p. 56). The outcome of the competition is considered temporary and strategic until something better comes along. In compatible pluralism relations among alternative explanations are emphasized. In that scientific explanations are usually causal explanations, identifying the conditions that give rise to the phenomenon can become the dominant focus of investigation (see also the contributions to *Scientific Pluralism*, 2006, edited by Kellert, Longino and Walters).

In Part I, Palombo (2013a) presents the Neuropsychodynamic Model with the following elements: 1) the self as a complex adaptive system; 2) the three levels of analysis of the self (Skurky, 1990), which include neuropsychological, introspective and interpersonal perspectives; and 3) the psychoanalytic glue holding the framework together is self psychology with narrative theory. A deficit theory explains psychopathology. Systems theory, adaptation and evolution clarify the grounding dimensions. Reminiscent of Freud's monumental Psychosexual Synthesis (Palombo, Bendicsen and Koch, 2009, pp. 24–33), Palombo trisects each of the three levels of analysis into developmental, dysfunctional states and treatment divisions. Of special interest is the section on separateness and individuality and his distinction among connectedness, differentiation and individuality which helps to clarify current thinking on attachment. (I discuss this topic above in the Attachment section.)

In Part II, Palombo (2013b) deepens the argument for the self as a complex adaptive system expanding the levels of analysis as a heuristic toward that end. The philosophical debate about the position of the observer in relation to that which is observed is considered. Positivists view the observer as independent of the observed and so reality is measurable, predictable and therefore knowable, whereas the constructivists hold to the perspective that the observer inexorably influences that which is observed and so reality is subjective, construed by both the observer and the observed, and cannot be known. There is no resolution possible to these viewpoints on reality, one empirical and the other hermeneutic. A way to sidestep this impasse is to consider the location of the observer in one of the three levels of analysis. Using a spacial metaphor, the observer stands on a platform where it is feasible to imagine three possibilities: at the neuropsychological level, the observer is located outside the system and is decontextualized from the observed (the extrospective perspective); at the psychological level, the observer is located within the system and joins or merges with the observed (the introspective level); and at the interpersonal level, the observer is located between the two systems, interacts with both and becomes a participant observer (the interactive level). Case examples illustrate that data may be gathered within a particular level and/or amongst levels enabling the estimation of a causal relationship or an association among variables. This flexible platform methodology facilitates the conceptual maneuverability of the self as a complex adaptive system.

In Part III, Palombo (2016) returns to his project of the revised contemporary human developmental model

> and proposes a set of processes that are consistent with a nonlinear dynamic perspective that govern human beings' mental activities during development. Three trends are identified, those that revised traditional theories of development, those that integrated the neurosciences with psychoanalysis/psychotherapy and those that integrated nonlinear systems theory into human development. These processes are the progression from lower to higher levels of complexity, the movement from lesser to greater differentiation and advancement from lesser to greater individuation.
>
> (p. 1)

Maturation of

> the self from a complex adaptive system is understood from either a developmental perspective (i.e., a historical viewpoint, the person is a product of his or her past) or from a dynamic viewpoint (i.e., in its contemporaneous state, where the person is a system of interactive components).
>
> (p. 2)

Contributions are reviewed from the first trend: Bowlby (1969; 1973; 1980), who, borrowing from the ethologists and the evolutionists, proposed that attachment was a species-specific behavioral pattern that infants manifest in the service of survival; Stern (1985), who proposed an alternative to the traditional developmental stages of the infant as reconstructed from adults in analysis, gathered empirical data and created a five-part theory of subjectivity where domains that last the life cycle replace periodic stages.

From the second trend: Schore (2001a; 2001b; 2011), who laid the neuropsychological foundation for attachment, postulated that the interaction between the infant's right brain with the caregiver's right brain generated an affect regulatory process leading to the infant auto-regulating affect states; Cozzolino (2014), who described the neurobiological processes in the social brain that regulate attachment and social relationships, created the term "social synapse" to describe the interaction of people essentially at the border of the synapse because most of the communication is out of awareness.

From the third trend: Palombo (1996; 2001; 2006), who had been studying dysfunctional states that result from learning disorders, focused efforts on the integration of concepts from self psychology with the neurobiological deficits found in learning disordered children; and Sander (2000; 2002), who integrated data from infant research, proposed that understanding mental health requires a multidisciplinary approach. "Sander viewed development as occurring within an evolutionary context that for human beings reaches its apex with the organization of consciousness and self-awareness" (Sander, 2000, p. 5, in Palombo, 2016, p. 8).

Sanders emphasized the need for specificity of fit between infant and caregiver, recognizing the wide cultural diversity in child-rearing practices. The degree of fit depends on the recognition process, understood as "an evolutionary construct through which adaptation occurs" (in Palombo, 2016, p. 8). Recognition is vital in that by being recognized the infant feels capable of affecting its environment. Recognition plays a significant role in developing a sense of agency and promoting a sense of coherence. These achievements contribute towards placing the self in a functional system resulting in a more differentiated hierarchical level of mental organization and complexity.

Palombo concludes Part III with a review of processes which govern the self as a complex adaptive system: 1) the elements in the system are not centrally controlled; 2) the elements are responsive to internal and external states in which the system exists; 3) a set of patterns form a map between the system and the environment such that the map acts as a guide to future interactions; 4) the complex adaptive system is a dynamic open system that is continuously changing and reorganizing itself. Growth is understood as an enlargement in system complexity, an increase in intrapsychic differentiation and an expansion in social individuation.

In Part IV, Palombo (2017a) turns to issues associated with a dynamic view of the self. Palombo states that the word "self" designates the mental activity of a person. It is grounded in an evolutionary framework of survival requiring successful adaptation. Adaptation occurs through the recognition process. It is understood through mindsharing, the psychological equivalent of recognition, as the central activity of the self. The nature of the self is considered first through an elaboration of the adaptation process. The process includes matching capacities between infant and caregiver through a subjective awareness of the success or failure of that matching: "Adaptation brings all the elements at play within the system into an integrated, organized coherent experience" (p. 7). Survival from physical and psychological perspectives may be at odds. Being heirs to our ancestral gene pool requires that we consider the success or failure of goals we set for ourselves.

Palombo next considers mindsharing as a central attribute of our sense of self. Mindsharing has two definitions: 1) a set of phenomenon in which one person can understand what is on another persons' mind, as in the capacity for empathy and 2) a set of functions, complementary in nature, in which each person is the recipient as well as the provider of such functions. Mindsharing, as a capacity, evolves to equip us with the capacity for interconnectedness and intimacy.

Palombo concludes with a discussion of the attributes of the self from the three levels of analysis: the neuropsychological, the introspective and the interpersonal. At the neurophysiological level essential elements are organized around innate/endowment (i.e., preset structures or patterns governed by the genome), those of mental functions embracing cognitive, affective and social domains. These properties allow for specificity of fittedness to account for our individuality and, alternately, blend in infinite ways accounting for our diversity. At the introspective level, "we are driven by two psychological biases or preference that contribute to our sense of self: *the preference for the subjective experience of self-cohesion* and *the wish for a coherent level of self-understanding or self-knowledge*" (Palombo, 2017a, p. 13). At the interpersonal level, "Most relational patterns are encoded in procedural memory" (Ogden, 2009, in Palombo, 2017a, p. 18). In addition to affective memory, it is when we have available episodic memory that we can develop narrative structures. Our capacity to dialogue with others is an essential attribute as we strive to reach out to others in interconnectedness.

Palombo has undertaken a significant project as he strives to reconfigure a psychoanalytic developmental model based more on empirical data than data reconstructed from patients' regressed states. This departure from traditional psychoanalytic oriented developmental orthodoxy opens the gate for a fuller interdisciplinary level of cooperation. Palombo's four-part project is a synthesis of available wisdom on the subject which continues Palombo's interest in recasting psychoanalytic concepts (1988) to make them relevant for today's clinician. The self as a complex adaptive system contains the dimensions of self psychology, relational theory, neurobiological research findings, attachment theory, evolution, narrative theory, mind sharing and complexity theory which also fit comfortably into the developmental algorithm project I am advancing.

Contemporary psychoanalytic developmental theory is indebted to the work of Greenspan and Shanker's (2004) life span developmental model. The Greenspan and Shanker framework is grounded in an evolutionary context that is sympathetic to non-linear developmental dynamics and is tied to attachment theory. Philosophically, it is non-teleological and non-deterministic. It is informed by neurobiological research findings, the study of autism in children and current infant observational data (e.g., the co-regulation of emotional communication). Of significance, the framework privileges the interactive exchange of affects between caregiver and infant as the origin of communication. The co-regulation of emotional signaling organized through the gradual differentiation of dual coding (the blanket is both smooth and pleasant) of experience provides the key to understanding how emotions organize symbol formation, intellectual abilities and, indeed, create the sense of self (p. 56). Two developmental researchers, Lichtenberg and Panksepp, have developed motivational models that rely on the primacy of affect regulation. Lichtenberg (1989; 2011), more clinically oriented, has argued for abandoning the drive model for a framework that comports more with emerging knowledge on the concept of motivation, including complexity theory, affect theory and caregiving. Panksepp (1998), more laboratory oriented, has formulated a neurobiological hypothesis suggesting that in the higher mammals there is an emotional operational system common to all which, in the case of humans, supports cognition, language, conscious and unconscious processes. In considering contemporary psychoanalytic developmental theory let us look at the contributions of Lichtenberg and Panksepp more in depth.

Lichtenberg and Hadley (1989), and later Lichtenberg, Lachmann and Fosshage (2011), based on extensive infant observations, eventually postulated that there were seven discreet motivational systems that must be understood in order to fully understand human behavior. Each of these systems is based on innate needs coupled with associated patterns of response. They are: the need for attachment to individuals; for affiliation with groups; for mutual psychic regulation of early physiological requirements such as hunger, elimination and sleep; for assertion, curiosity, and exploration of preferences and capacities; for reaction to aversive experiences through withdrawal and/or antagonism; and for sensual enjoyment and, ultimately, sexual excitement (Gabbard, 2014, p. 56; Lichtenberg and Hadley, 1989, p. 372; and Lichtenberg, Lachmann and Fosshage, 2011, p. 30).

> Each system self-organizes and self-stabilizes as a loose assembly of categorized experiences having similar but not identical affects and purposes. A predominant similarity of affect, intention, and goal provides the basis for our proposal of each motivational system as a conceptual entity. Once self-organized and self-stabilized, each system remains in dialectal tension with other intentions and goals within the same system, with other systems of the individual, and with convergent and divergent intentions and goals that arise from immersion in an intersubjective matrix. Dialectic tension can result in activation or deactivation of dominance of the individual's mental state by one or another motivational system. The shifting of dominance generally proceeds smoothly, often without notice, and usually without any alteration in the sense of self-identity.
>
> (Lichtenberg, Lachmann and Fosshage, 2011, p. 30)

It is assumed that each motivational system has aims and goals maintained by the memory system and mature epigenetically (Gedo, 1988).

Panksepp, drawing data from neurobehavioral knowledge, especially neuroanatomy, neurophysiology and neurochemistry, constructed a theory of the fundamental emotional tendencies of all mammals.

> To make mechanistic sense of certain brain systems in humans, we must imagine how the simplicity of certain animal behaviors relates to the vast complexity of human experience, and vice versa. This is certainly a hazardous undertaking, but … I will pursue the idea that the mammalian brain contains a "foraging/exploration/investigation/curiosity/interest/expectancy/SEEKING" system

that leads organisms to eagerly pursue the fruits of their environment – from nuts to knowledge, so to speak. Like other emotional systems, arousal of the SEEKING system has a characteristic feeling tone – a psychic energization that is difficult to describe but is akin to that invigorated feeling of anticipation that we experience when we actively seek thrills and other rewards. Clearly this type of feeling contributes to many distinct aspects of our engagement with the world.

(Panksepp, 1998, p. 145)

The central circuits for this central brain function are concentrated in the extended lateral hypothalamic (LH) corridor. This system responds unconditionally to homeostatic imbalances (i.e., bodily need states) and environmental incentives. It spontaneously learns about environmental events that predict resources via poorly understood reinforcement processes ... I believe these transhypothalamic circuits lie at the very heart of the SEEKING system.

(p. 145)

Panksepp (1998; 2010; 2012), based on research with laboratory rats, formulated a theory of motivation linked to a set of seven distinct, but often overlapping, emotional systems that emerged during evolution to serve adaptive functions. "The emergence of emotional circuits, and hence emotional states, provide powerful brain attractors for synchronizing various neural events so as to coordinate specific cognitive and behavioral tendencies in response to archetypal survival problems" (Panksepp, 1998, pp. 303–304).

Each of these systems is affectively valenced, yielding feelings that are either positive or negative, desirable or undesirable, but there are probably several distinct forms of each of these general types of affective experiences. Considerable evolutionary diversity has been added by species-typical specializations in higher brain areas as well as lower sensory and motor systems, but as we have seen, the basic affective value systems, deep within ancient recesses of the brain, appear to be reasonably well conserved across mammalian species.

(Panksepp, 1998, pp. 303–304)

Panksepp also has important things to say about the origin of the Self.

Panksepp's archaic self is a biological notion of identity. It is a concept of self based more on affectively rich *action* rather than rarefied intellectual *reflection*, and so it includes many other kinds of non-human animals in the club of selves ... Affective neuroscience reminds us of our phylogenetic homologies with other mammals, and so our biological identity should be found near the core of the brain – not the more recent cortex. This archaic Self would be a basic motor-mapping system – a template for action tendencies. Despite

TABLE 5.1 Panksepp's theory of motivation

Emotional system	Subjective experiential state(s)	Behavioral pattern(s)/intentionality
Seeking	Curiosity	Approach
Lust	Sensuality	Sexual posturing
Fear	Trepidation	Escape
Rage	Anger	Attack
Care	Affection	Seeking nurturance and support
Panic	Sorrow/grief	Withdrawal
Play	Joy	Exuberant activity

(Chart adapted from Panksepp, 1998 and Virginia Barry, conference presentation on February 3, 2018 and personal communication on April 4, 2018)

the inclination of philosophers to think about consciousness and subjectivity in terms of perceptions (like sense data qualia), affective neuroscience reminds us that "a level of motor coherence had to exist before there would be utility for sensory guidance" … the organism is establishing attraction and aversion values at the subcortical level, and so the organisms most rudimentary self-awareness, of a spatio-temporally located body in an environment, will already be coded with positive and negative affects. The self is not superadded after a certain level of cognitive sophistication is achieved (a view commonly held by philosophers). Rather, the self first emerges in the precognitive ability of most organisms to operate from an egocentric point of view. Way below the level of propositional beliefs, animals must solve basic motor challenges.

(Panksepp, et al., 2010, pp. 29–30)

So Panksepp, informed by neurobiological research, locates a theory of motivation, as well as a theory of the origins of the self, deep in core brain biological motor activity, minimizing proto-cognitive processes. Rather than Descartes', "I think, therefore I am," Panksepp suggests "I feel, therefore, I am" (Panksepp, 1998). Panksepp joins company with the embodied metaphor researchers, such as Lakoff, and embodied philosophers, such as Merleau-Ponty, in locating the origin of human knowing and functionality in biological processes.

Both the Panksepp and Lichtenberg and Hadley et al. hypotheses are data-driven contemporary motivational systems that serve to supplement Kohut's (1959; 1966; 1971) theory of healthy narcissism and narcissistic strivings. These compelling frameworks threaten to supplant Freud's once dominant dual-instinct drive theory of libido and aggression, the life and death instincts (Freud, 1920, pp. 34–64). With compelling findings from infant observation and neurobiological research, the time has come to consider a more nuanced, integrative approach to human motivation. These motivational systems compliment the emerging framework of a contemporary psychoanalytic developmental theory that enables us to move beyond classical instinctual theory. Both frameworks are grounded in evolutionary biology, informed by infant and mammalian observation and neurobiology and the respective states/stages appear to unfold in probabilistic epigenetic sequences which enhances their contributions to the contemporary psychoanalytic developmental model. In Lichtenberg's case his framework is grounded decidedly in psychoanalytic conceptualizations. Of course, to Lichtenberg and Hadley, Lichtenberg, Lachmann and Fosshage and Panksepp's theories of motivation must be added that of Cozolino (2006) and his neurobiological Social Brain theory of motivation. See Chapter Two for details.

In order to comport regulation theory into my developmental algorithm, I have translocated Greenspan and Shanker's model into a modification to Kohut's (1977, p. 97; 1991) division of the self-structure into subordinate (more empirical) and supraordinate (more abstract) dimensions. I have formulated a substitution of Greenspan and Shanker's model for Mahler's separation-individuation framework as the subordinate component linking Kohut's supraordinate metapsychology of the self for a more efficacious, contemporary, cross-disciplinary perspective (Palombo, Bendicsen and Koch, 2009, p. 263; Bendicsen, 2013, pp. 59–70; 180–182).

Developmental processes

Contemporary psychoanalytic developmental theory also embraces two opposing viewpoints. The far view, the view from above (in the philosophically modern, traditional, positivism orientation), is the group perspective. It embraces a developmental model grounded in linear systems theory in which development marches forward in sequential, invariant steps. The near view, the view from below (in the more philosophically postmodern, nontraditional, social constructivism perspective), emerging from non-linear dynamic systems theory, is the individual position. It recognizes and accommodates process, over steps, in the infinite variations of individual growth.

The second perspective, the near view, allows us to move away from understanding growth and progress as movement along preset steps to a model that detects growth and progress as movement and shifts in states that may lack traditional forms of knowing. "In dynamic terminology, then, behavioral development may thus be envisioned as sequences of system *attractors* of varying stability, evolving and dissolving over time" (Thelen and Smith, 1994, p. 86). Patterns emerge exclusively as a result of component cooperation. One pattern may dominate over others only to change in time: "these notions of evolving and dissolving attractors, representing various states of cohesion of the components, will apply equally well for real-time cognition and new developmental forms" (p. 86). Individual growth and progress is understood more as randomized differentiation in shifting self-states which do not conform to traditional patterns and pathways.[1]

As mentioned in the Preface, the focus of this monologue is on explicating the dynamics associated with the developmental passage from late adolescence into young adulthood. Traditionally, the passage was assessed as the measure by which a collection of life tasks were addressed and completed. Despite advances in knowledge, the construct of life tasks has a remarkable adhesiveness among the lay public. Professionals, however, are now trained to look to other criteria by which to measure this passage. I will shift the discussion from tasks to be accomplished to the set of processes that need to be understood. Let us consider the growth processes available to the emerging adult organized into biological and then psychological forces.

A. First are those associated more with evolutionary/biological maturational forces that signal advanced levels of differentiation and reorganization:

In addition to the need for 1) attachment-differentiation/individuation as the emerging adult utilizes the reconfiguring larger self-selfobject milieu to further differentiation, and 2) the working through of cognitive dissonance oriented in active (reflexive) vs passive (reflective) neurobiological tension to achieve the capacity to delay gratification, the hallmark of self-regulation in maturity (Steinberg, 2014, pp. 107–124), there are other developmental forces at play in opening the gateway into young adulthood.

3) The central point I made in the *Transformational Self: Attachment and the End of the Adolescent Phase* was that adolescence can be said to "end" with the instantiation or activation of the Transformational Self. Organized around the appearance of the self-referencing metaphor, the new set of attendant identifications, possibilities and capacities coalesce, furthering developmental processes specific to the transmutation of the Transformational Self into the psychology of the emerging adult needs to take place. They include first integrating the new, phase-specific neural self into a consistent way of behaving and thinking about the reconfiguration. The self-referencing metaphor then evolves into a self-regulating metaphor that channelizes experience, creating a consistent, stabilizing focus of change. This is the work of the self-selfobject milieu, a support system that serves to consolidate the reconfigured self.

4) Next, the raw grandiosity linked to the new self constitutes the affective energy that propels development. It disturbs the cohesion of the self in a manner similar to that which happened to the latency character structure encountering puberty. Weakening and abandoning old identifications leaves the adolescent feeling a need to test out and publicly exhibit new capacities while still in a state of vulnerability and ambiguity. This grandiosity needs to undergo transmutation (the internalization of selfobject functions) in the development of new internal multiple psychic structures so that it stabilizes and contributes to the reconfigured self a sense of vitality, exuberance and excitement in the evolving narration of possibility.

Let us explore further the ways in which grandiosity undergoes modification from its mobilization in phase-specific adolescent developmental exhibitionism to a regulatory feature of self-esteem? After Myles secured a new position as merchandizing and pricing associate he encountered substantial frustration in adjusting to the computerized flow of itemized merchandise through the department. "The 'Top dog on the floor' has met his match," he said. "I am running the risk of being ordinary, even subpar because I can't keep up. I'm losing focus and have become easily distracted. My supervisor notices and said I need to finish one task completely before moving on to the next." Myles' solution is to ask his former peers, retail associates, for help during slow periods on the floor. This solution is a customary one but requires good collaborative relations with colleagues who, strictly speaking, do not have to help him. The recognition he enjoyed as a top performer carried with it fantasies of entitlement, superiority and imperiousness. These aspects were not acted out and did not materialize because Myles enjoyed helping new employees learn the ropes and was the "go to" guy on the floor when issues arose. Myles was calling in his markers, always being aware of the three great vulnerabilities under stress that could doom his future: his impaired self-assessment (especially personal hygiene), his impulsive communications and his tendency toward dysregulation.

There are at least three selfobject configurations which will help me explain how grandiosity gets attenuated:

a) "*alter-ego* selfobjects sustain the self by providing the experience of a perceptible presence of essential likeness of another's self" (Wolf, 1988, p. 185);
b) "*adversarial* selfobjects sustain the self by providing the experience of a center of initiative through permitting nondestructive oppositional self-assertiveness" (p. 185);
c) self-delineating selfobjects sustain the self by a process in which the self acquires an experience of the world and the self as real. "'Reality,' as we use the term, refers to something subjective, something felt or sensed, rather than to an external realm of being existing independently of the human subject. (Stolorow and Atwood, 1992, pp. 26–27).

> It is our view that the development of the child's sense of the real occurs not primarily as a result of frustration and disappointment, but rather through the validating attunement of the caregiving surround, an attunement provided across a whole spectrum of affectively intense, positive and negative experiences. Reality thus crystalizes at the interface of interacting, affectively attuned subjectivities.
>
> (p. 27)

> The self-delineating selfobject function may be pictured along a developmental continuum, from early sensorimotor forms of validation occurring in the preverbal transactions between infant and caregiver, to later processes of validation that take place increasingly through symbolic communication and involve the child's awareness of others as separate centers of subjectivity.
> (Stolorow and Atwood, 1992, p. 27 and also pp. 35, 49, 70, 82, 95)

In so far as the late adolescent/young adult is concerned, the alter-ego, adversarial and self-delineating selfobject functions exert a significant force in this phase because they concern new subjective interactions outside of the family in the environmental encounter with peers and authority figures. In Myles' case, respectively, grandiosity is attenuated:

1) for the alter-ego selfobject function through interaction in a cohort/team striving to achieve group, not just individual goals;
2) for the adversarial selfobject function through dealing with authority figures in the world of work who generate stress and tension by criticizing/critiquing work performance, especially the subpar performance of underachieving coworkers; and
3) for the self-delineating selfobject function through employment vicissitudes, including performance evaluation and gauging promotion potential, where the cohesion of the vulnerable self is threatened.

In the average expectable job environment, opportunities for repair emerge allowing these three types of self-selfobject encounters to function as opportunities in affectively attuned circumstances to "articulate and consolidate subjective reality" (Stolorow and Atwood, 1992, p. 95) and to strengthen resilience of the self-state.

The aforementioned grandiosity seems consistent with the frequently observed egocentric phenomenon considered "to be fairly universal in adolescence in our culture (Elkind and Bowen, 1979):

a. Feelings adolescents have of being in front of an imaginary audience before whom they are performing.
b. A belief in a personal fable that gives expression to the sense of uniqueness and specialness adolescents feel.
c. The experience of what has been described as a 'time warp.' Adolescents are unable to see themselves in a historical perspective in which the present relates to the past and the future.
d. The manifestation of a 'cognitive conceit.' Adolescents believe they know more than their parents or than any other adults. They have an absolute certainty of being right."

(Palombo, 1988, pp. 174–175)

These dimensions of grandiose-egocentricity capture the sense of unlimited possibilities when late adolescents mobilize the self-referencing metaphor.

5) There is an additional process that needs to be highlighted. Is there a way to think about the consolidation of the young adult character structure that further clarifies the process? It seems to me that as the Transformational Self is mobilized, in addition to attenuating the attendant grandiosity another process unfolds, that of ambivalence. As the grandiosity subsides, a sense of vulnerability ensues with the emerging adult pondering this new state of affairs. We can hear the emerging adult wondering, "Is this really me? How can I trust this new way of being in the world?" A cognitive feeling of normative dissonance emerges understood as a discrepancy between the old and new selves. Harter considers this a normal phenomenon as the

adolescents' discrepant self-representations struggle to achieve self-coherence and self-continuity (Harter, 2012, pp. 96–98). Normative dissonance is often associated with the study of late age development. Colarusso (1992) employs normative dissonance as an aspect of late life cognition and defines it as "a discrepancy between the intrapsychic sense of the body and the sentient self, the latter being experienced as younger and more vigorous, imprisoned, in a sense, in a shell of a body which is no longer compatible with the mind, or able to carry out its commands" (p. 184). However, the term normative dissonance has conceptual usefulness throughout all developmental transitions as a way to capture and describe the continual cognitive/affective reconfiguring of the self-state.

How do we know reality? I suggest we start with Sullivan's term, consensual validation (1953, p. 29; 1962, pp. 258–259; 1964, pp. 163–164), defined as a communication process whereby meanings about verbal and nonverbal perceptions and concerns are mutually agreed upon by two or more persons resulting in a confirmation of reality including modifications and elaborations of distortions. Building on this definition, two highly useful conceptualizations have emerged to describe the challenge of resolving the normative ambivalence associated with consolidating the old and new selves. How does our lived experience present us with, and confirm for us, what is real?

The first concept comes from advances in attachment theory, regulation theory and dynamic systems theory. Tronick (2007) has formulated the MRM to capture the importance of the infant-mother dyadic interaction as one of co-regulation and co-created meaning-making that utilizes energy and new information to generate a sense of being-in-the-world. As the ever reconfiguring self expands, communication through the exchange of affects organizes development according to the now well-known dynamics of complex systems (Siegel, 1999, pp. 208–238). As dyadic states of consciousness unfold, a sense of joyful expansion through activity is experienced which creates a sense of continuity and certitude about one's place in the environment. With this movement into the world the individual realizes an awareness that "I know this to be true."

The second concept originates from the intersubjective work of Stolorow and Atwood (1992, pp. 26–27) and their expansion of the role of the selfobject, specifically the self-delineating selfobject. As the self acquires experience of being-in-the-world, a sense of subjective reality is felt or sensed. This is not a sense of independence or autonomy. Rather it is a sense of reality acquired through validating attunement in which two or more subjectivities exchange affective communication in such a manner that the contours of the self are defined and delineated. These contours of the self are reinforced through participation in the selfobject milieu through innumerable caregiving interactions. As the adolescent differentiates, the self-delineating selfobjects help reshape the sense of reality through a working through of the phase-specific normative dissonance leading to awareness of the reconfigured self as a separate center of subjectivity.

Second are those associated more with philosophical/cultural developmental forces that serve as imperatives in the creation of the capacity for resilience (defined as successful adaptation in the face of adversity) necessary for healthy functionality in contemporary society:

6) Summers (2013) has recently corrected the traditional emphasis on the past and present in psychoanalytic work by stressing the role of futurity. Summers, citing the work of philosophers, believes that

> life is lived in the future perfect tense, and consequently, the meaning of life events has the character of "will have been done." The meaning of any action lies in what is portended, conceived as completed in the future. No experience in the present can have any meaning without its intentionality, the aim toward which it is directed. The present moment derives its meaning from how it fits into the plan of action, however simplistic and implicit it may be. The past becomes relevant in so far as it is encountered in the trajectory toward the future. The future then, is ontologically prior to both present and past.
>
> (p. 112)[2]

7) In addition to understanding the future as an ontological force pulling the late adolescent into young adulthood, we need to consider the domain of intersubjectivity and Benjamin's emphasis on the development of recognition (Benjamin, 1990). Benjamin combines concepts from Mahler, Pine and Bergman's (1975) separation-individuation theory, Winnicott's (1964, 1971) object relations theory, Stern's (1985) interpersonal framework and Atwood and Stolorow's (1984) intersubjectivity theory to arrive at her four-stage developmental hypothesis.[3] (Elaborations of these frameworks can be found in Palombo, Bendicsen and Koch, 2009.)

Benjamin's developmental trajectory of mutual recognition in intersubjectivity is a blend of intrapsychic and intersubjective dimensions. In her words:

> Its core feature is recognizing similarity of inner experience in tandem with difference. We could say it begins with "We are feeling this feeling," and then moves to "I know that you, who are another mind, share this same feeling." In rapprochement, however, a crisis occurs as the child begins to confront difference – "You and I don't want or feel the same thing." The initial response to this discovery is a breakdown of recognition between self and other. "I insist on my way, I refuse to recognize you, I begin to try to coerce you; and therefore I experience your refusal as a reversal: you are coercing me. Here the capacity for mutual recognition must stretch to accommodate the tension of difference, the knowledge of conflicted feelings."
>
> (Benjamin, 1990, pp. 42–42)

The differentiation process continues in stage four in the third year with symbolic play and

> symbolic understanding of feeling so that "You know what I feel, even when I want or feel the opposite of what you want or feel." This advance in

differentiation means that "We can share feelings without my fearing that my feelings are simply your feelings."

(Benjamin, 1990, p. 43; Summers, 2013, pp. 34–36)

I believe Benjamin implies that the differentiation of mutual recognition is a process for the lifespan. A continual tension exists between the self and other, on the one hand, by relating as subject and object in a reversible complementary relationship of alternating power and control rather than, on the other hand, as a balance of destruction and recognition between two subjectivities struggling for mutual understanding. It seems to me that as development unfolds, accompanied by different sets of potentialities, the tension between self and other needs to be renegotiated at every developmental juncture where self-regulation is achieved by regulating the other.

These seven developmental processes are:

A. Those associated more with evolutionary/biological maturational forces that signal advanced levels of differentiation and reorganization:

1) the need for attachment (interconnectedness) – differentiation/individuation (Lyons-Ruth, 1991; Doctors, 2000; Palombo, 2016);
2) the need for cognitive dissonance to resolve opposing, conflicting biological maturational states into one coherent schema (Bendicsen, 2013);
3) the mobilization of the Transformational Self exerting the force of a dynamical attractor (Bendicsen, 2013);
4) the attenuation of the sense of grandiosity that accompanies the Transformational Self (Bendicsen, 2018); and
5) the need to consolidate the reconfigured self-state with a sense of certitude (Tronick, 2007) in a new reality about being a distinctive subjectivity in an intersubjective environment (Stolorow and Atwood, 1992).

B. Those associated more with philosophical/cultural developmental forces that serve as imperatives in the creation of the capacity for resilience (defined as successful adaptation in the face of adversity) necessary for healthy functionality in contemporary society.

6) the desirability of living a life organized into the future (Summers, 2013);
7) the differentiation of mutual recognition between two subjectivities, self and other (Benjamin, 1990), which is a process for the lifespan that needs to be renegotiated at every developmental juncture.

These seven processes exert a dynamism (from Sullivan, 1953, p. 103, "the relatively enduring pattern of energy transformations which recurrently characterize the organism in its duration as a living organism," in Palombo, Bendicsen and Koch, 2009, pp. 209, 231) of action potential, an ontological force, that collectively contributes to pulling the late adolescent into young adulthood. These processes unfold neither sequentially nor simultaneously, but rather manifest according to organizational dynamics commensurate with complexity theory.

Complexity theory or non-linear dynamic systems theory

Complex systems refers to a

> scientific field that studies the common properties of systems considered complex in nature, society and science. Complex systems is often used as a broad term encompassing a research approach to problems in many diverse disciplines including neuroscience, social science, meteorology, chemistry, physics, psychology economics, earthquake prediction, molecular biology and inquiries into the nature of living cells themselves.

Alternative terminology for complex systems includes dynamical systems, non-linear systems, chaos theory and complex systems theory (Charles Jaffe, handouts from conference presentation on February 3, 2018).

The emergence of the self-referencing metaphor in late adolescence heralds the development of the beginning formation of the ego/self ideal. The ideal self is a reconfigured self-state or, in dynamical systems terminology, an attractor state shaping subsequent development. "An attractor is a pattern of motion toward which a system tends" (Galatzer-Levy, 2016, p. 412). At this point let us turn to Siegel (1999) and his ideas on dynamic systems and the linkage with neuroscience research findings. Three features define a dynamical system: 1) they have self-organizing properties; 2) they are non-linear; and 3) they have emergent patterns with recursive characteristics. Self-organizing properties involve the notion of the development of each human being moving from simplicity toward complexity. Continuous movement toward maximum complexity promotes system stability, understood here as optimal neuronal connectivity. The strength of synaptic connectivity is altered by experience. Repeated mobilizations of a particular profile of activations, a state of mind, can make such a configuration of neuronal assemblies a deeply engrained attractor state. An attractor state is a stable pattern of environmental interactions, patterns of coordination that are never identical, occurring in a specific context. Non-linearity refers to the idea that system output is context dependent and, therefore, unpredictable. In other words, a small change in input (such as alterations in one's beliefs, emotions and perspectives) can lead to disproportionately large behavioral changes. "'Emergent' means that each of us is filled with a flow of states that evolve across time. 'Recursive' means that the effects of the elements of a given state return to further influence the emergence of the state of mind." Taken together, emergent and recursive refer to the principle of change being self-perpetuating, continuous and moving toward differentiation and ever new states of integration. The self as a dynamic adaptive system is always in a state of construction and reconstruction (pp. 217–222).

Jaffe (2018) has crafted a developmental view of adolescence from a dynamic systems configuration.

> Within a dynamic systems perspective, adolescence may be considered a period of reorganization in the context of greater complexity. In the most generic sense, the shift occurs simply because contexts for adaptation change. We call these contexts for adaptation the developmental tasks of adolescence. But there is no inborn, supraordinate, primary motive such as pressure towards instinctual discharge, or a need to separate from infantile objects, or a new selfobject requirement. Instead, the adolescent process may be considered in terms of the shifting equilibrium of the organismic system, or self. It occurs as it does, in various forms in history and cultures, because the subsystem assemblies that are necessary for adaptation vary in context. Some integration will occur, but the on-the-fly, opportunistic, context sensitive nature of adaptation means that this will not be the same across all time and cultures. In other words, one would expect that some new integration, or identity will emerge, but its behavioral manifestations, and even the individual's subjective sense of it, will always reflect the context specific nature of its occurrence. Developmental tasks are clearly not the same for everyone. For example, one can imagine that an individual's identity within a caste system, where all aspects of life are a function of specific status, would be quite different than a sense of oneself in a system where one's role and status shifted as one moved in different areas of life (Samereroff, 1983).
>
> (Jaffe, 2018, p. 19)

Palombo (2013b; 2017b) has formulated a developmental theory based on principles from, among other theories, self psychology, neuroscience research findings and core concepts from dynamical systems theory. Following Skurky (1990), he constructs an observational platform consisting of three levels of data observation and collection: the neuropsychological level, an introspective level and an interpersonal level. His quest is to fashion a unified psychoanalytic clinical paradigm that would bring together psychoanalytic and nonpsychoanalytic theories. This example is but the latest attempt to combine information from disparate knowledge domains to create a more satisfying explanatory framework to categorize human experience. In this regard Palombo's project comports well with my developmental algorithm.

Galatzer-Levy (2016; 2017) has formulated a set of techniques using complexity theory. Five concepts from complexity theory are discussed: the edge of chaos, emergence, attractors, self-similarity and coupled oscillators. The edge of chaos refers to the point at which change occurs in the system. The edge of chaos is the reality of emergence, a shift in system properties. It is not so much a shift in direction (a linear term), but rather a reconfiguration of elements such that a new level of differentiation is reached. Emergence is understood to mean "the phenomenon of surprising novelty that occurs when elements are placed together in particular contexts" (Galatzer-Levy, 2016, p. 421). "Qualitatively, these new system aspects are not reducible to the sum of the system's elements, rather the system as a whole must be considered. An attractor is a pattern of motion toward which the system tends" (p. 412).

> An attractor can be thought of as a path along which a person travels with multiple offshoots or directions. Self-similarity refers to the phenomenon in which examining a small dimension of the attractor reveals information about the entire attractor, so that altering the small segment impacts the entire system in a non-proportional way. Oscillators are entities that move in a range between end points.
>
> (p. 415)

When two oscillators are joined (as in two side-by-side pendulums hung on the same wall), a new oscillator is created with richer capacities than either oscillator alone or when both are added.

Galatzer-Levy (2016; 2017) suggests that change in psychoanalytic treatment occurs at the edge of chaos with new phenomenon in a differentiating reality of emergence influencing an attractor that shapes subsequent development. The shift in the system's nature can involve a small perturbation amplifying forces, in humans that are often communication related, with disproportional effects. For Myles, chaos was the schizophrenic break, the terrifying experience of the sun bursting (Bollas, 2015). The edge of chaos can be thought of as recovery, the collective impact of the treatment plan and especially the emergence of the powerful attractor, the self-referencing metaphor. The interaction between patient and therapist can be understood as a coupled oscillator mutually influencing each other and subsequently the course of treatment. We labeled Myles' illness as "the condition." In objectifying the schizophrenic illness, its terrifying, out of control, affective firestorms underwent a kind of conversion into a topic of conversation. With the emergence of a co-constructed narrative the alien third person condition was on its way to becoming integrated. The self-similarity became the perception that the condition, as long as the treatment plan was followed, was more under Myles' control. With the reduction in symptoms a sense of personal agency emerged resulting in a reconfigured stable self-state. When Myles uses the "Top dog on the floor" self-referencing metaphor, a new self-state, a different state of mind, is created. Myles' chaotic life experienced a novel realignment of variables such that the force of an attractor emerged organizing thoughts about identifications and their following associations into a set of fresh possibilities.

The application of complexity theory to single-case examination compels psychoanalysis to move into a multidisciplinary context. Complexity theory, a mathematical, nonpsychoanalytic construct, can be linked with psychoanalysis to enhance understanding. In my view, the resulting explanatory synergy confirms the usefulness of the plurality of theoretical orientations.

Neurobiology with narrative theory

Because the Transformational Self concept emerges from neurobiological research activity, let me briefly discuss its origins. I define the neural self using three sources. Damasio (1994, pp. 236–244) defines "The neural self, as a complex adaptive system, is a repeatedly reconstructed biological state that endows our experience with subjectivity and that depends on the continuous reactivation of images about our identity and our body." From Schore (2002, pp. 443–448) we borrow: "In addition, the neural self emerges as a result of right hemisphere maturation. It is a body self through the gradual differentiation of mutual co-regulation of state and affective experience between the caregiver and the self." Feinberg (2009, p. xi) defines the neural self "essentially by its coherence: *a unity of consciousness in perception and action that persists in time*" (in Bendicsen, 2013, pp. 185–186). Also, the self of self psychology breathes metaphoric life into the neural self. The neural self, then, incorporates dynamic systems thinking, embodied subjective processes, mutual co-regulation among other neural selves and a measure of its tonus or vitality through evaluating its coherence. The Transformational Self (Bendicsen, 2013) is a phase-specific neural self which can become mobilized with the activation of the self-referencing metaphor at or near the end of adolescence.

Myles' "condition" can be understood as a case of biological dysregulation involving hemispheric imbalance. Let us see how this hypothesis might work. Neurobiological research findings and narrative theory are joined to underscore the profound influence that narrative has on neural networks. Cozolino (2010) links neural networks and narratives into a quest for multilevel integration of the self. Two information flows relevant to psychotherapy have been discerned to be central in the study of self integration: "top-down (cortical to subcortical and back again) and left-right (across the two halves of the cortex)" (p. 27).

> Due to the interconnectivity between left-right and top-down neural networks, examining integration from either the vertical or horizontal dimension alone is overly simplistic. Studies of metabolic activity in specific areas of the brain in pathological states reveals differences in both cortical and subcortical structures on both sides of the brain. This research suggests that restoring neural integration requires the simultaneous reregulation of networks on both vertical and horizontal planes. It is also important to remember that although we are discussing brain functioning from the perspective of neural networks, an equally meaningful discussion could focus on the impact of pharmacological agents on the modulation and homeostatic balance of these same networks (Coplan and Lydiard, 1998). This perspective helps us to understand why both psychotherapy and medication can result in shifts of neural activity and symptom reduction and why together they may work better than either one alone (Andreasen, 2001).
>
> (Cozolino, 2010, pp. 28–29)

Narratives perform an array of functions including:

> Grounding our experiences in a linear sequential framework
> Remembering sequences of events and steps in problem solving
> Serving as blueprints for emotion, behavior, and identity
> Keeping goals in mind and establishing sequences of goal attainment
> Providing for affect regulation when under stress
> Allowing a context for movement to self-definition.
> (Cozolino, 2010, p. 163)

It is obvious that these narrative functions exert a compelling influence on the emerging, self-referencing metaphor of the late adolescent. With neuroscience research findings corroborating the influence of narratives on both the organization of and the trajectory of the new (neural) state of mind, the late adolescent is now in a position to harness the potential effect of prefrontal cortical maturation and its enhanced connectivity. This effect acts as an executive function force to design a life originating in a vision of self with new opportunity and possibilities.

It is helpful to remember that "our brains are designed to have a sense of causality between certain temporally related events and also to classify and categorize objects and events in certain ways" (Panksepp, 1998, p. 21). This biological imperative drives narrative formation. The degree of coherence in the narrative is the measure to which a state of mind guides the executive function of agency in the self.

From a neuro-biologically informed perspective, one can think of the self, more properly the neural self, as a state of mind. Following Siegel (1999), 1) "A 'state of mind' can be defined as the total pattern of activations in the brain at a particular moment of time" (p. 208); 2) "A state of mind ... involves a clustering of functionally synergistic processes that allow the mind as a whole to form a cohesive state of activity" (p. 209); 3) A state of mind "coordinates activities in the moment, and it creates a pattern of brain activation that can become more likely in the future" (p. 210). The neural self, then, is a coherent dynamism of mental processes, "a unity of consciousness in perception and action that persists in time" (Feinberg, 2009, p. xi). Metaphorically, the neural self is a stable personality organization of energy potentials constantly undergoing reconfiguration. For a fascinating discussion of the neural self from a neuro-biological perspective see Feinberg (2009, pp. 132–158).

Why do I link neurobiology with narrative processes? Cozolino (2010) emphasized that self-reflection offers a window into shifts in states of mind that reflect the activation and integration of different neural networks. These shifts come about through the interplay among different perspectives, emotional states and ways of using language. Three levels of language processing are explicated during the shifts in states of mind: a reflexive social language (RSL), an internal dialogue and a language of self-reflection. Reflexive social language, primarily a function of the left hemisphere, is an automatic stream of words, for example, clichés, that facilitates the ongoing communication of social relatedness. Internal dialogue, primarily a function of the right hemisphere, is a private communication between two aspects of the self, perhaps the experiencer and the observer as in a dream. It is shaped by personal emotions and may be used to deceive others.

> Like RSL, internal dialogue is primarily reflexive and based on semantic routines and habits reflecting our learned history. We hear in our heads the supportive or critical voiced our parents implanted early in our life. So while RSL keeps us in line with the group, internal dialogue keeps us in line based on early programming.
>
> (p. 170)

Self-reflection is less a vehicle for social control and more a mechanism for thoughtful consideration and potential change. "It employs executive function and serves to develop a theory of our own mind" (p. 171). While RSL and internal dialogue reflect unconscious aspects of the self, self-reflection may reflect a higher level of integration, deliberate, conscious processes which promote self-observation and self-evaluation. "In this language, cognition is blended with affect so that there can be feelings about thoughts and thoughts about feelings" (p. 171).

These languages serve to combine in the interweaving of narrative storytelling among therapist, client and others. In this language matrix an active editing process occurs, enabling co-construction of the self-narrative that can hold and reshape the self-referencing metaphor to facilitate forward progress, or rather, greater differentiation.

In other words, with the appearance of the self-referencing metaphor, the potential now exists for the self to organize around a new image of possibilities. This potential does not occur until late adolescence.[4,5] Why should this be the case? The key is the maturation of the prefrontal cortex and its enhanced neural assembly connectivity allowing for increased executive function capacity. Even a relatively minor change in self-image can result in an attractor state, an altered state of mind that constitutes a powerful new identity synergy. This image can pull the self forward into focused activities that are narcissistically invested with hope for a reconstituted identity. As mentioned earlier, self-referencing metaphor can evolve into self-regulating metaphor, stabilizing the new self-state through the maintenance of a set of coherent identifications, skill sets and enduring beliefs. As the evolving self narrative germinates in enhanced coherence, the reconfiguring self-state registers improved cohesiveness.[6] In an earlier publication I have labeled this reshaped identity the Transformational Self, a self-state of mind that can open the gateway into young adulthood (Bendicsen, 2013).

Myles, in associating to the "Top dog on the floor" metaphor, said, "I'm the best they got." Myles has understood the reality of the past, measured himself in the present and has located himself in the possibilities of the future. In addition, Myles' capacity for mutual recognition among a host of subjectivities is accompanied with a keen sense of differentiation among the ownership of feelings and those which are shared. If his "condition" deteriorates, this capacity will certainly worsen. The secure attachment he experienced as a child now serves as a vital underpinning for further growth. Myles will need every measure of his narrative strength as he struggles to stabilize and manage his "condition."

Notes

1 What is the difference between attractors and strange attractors?

> When systems self-organize under the influence of an order parameter, they "settle into" one or a few modes of behavior (which themselves may be quite complex) that the system prefers over all the possible modes. In dynamic terminology, this behavior mode is an attractor state, as the system – under certain conditions – has an affinity for that state. Again in dynamic terminology, the system prefers a certain topology in its state space.
>
> (Thelen and Smith, 1994, p. 56)

Strange attractors, also referred to as chaotic attractors, are extraordinarily complex on many different state spaces: "when plotted on appropriate state spaces, a highly complex, but deterministic pattern emerges, suggesting an underlying global order to the system behavior, but a local unpredictability" (p. 60).

Siegel (1999) links strange attractors to psychopathology, viz., disorganized and unresolved states of mind in attachment theory.

> For the person who has experienced disorganized attachment, the experience of parental fear or fear-inducing behavior has often been associated with parent's lack of resolution of trauma or loss. That is, the incoherence of the parent's life narrative has been behaviorally injected into the child's experience by way of the parent's own disturbance in self-organization and the resultant dysregulated states and disorienting actions. These parental behaviors, which are incompatible with providing a sense of safety and cohesion, are "biological paradoxes" and directly impair the developing child's affect regulation, shifts in states of mind, and integrative and narrative functions. The result is that the child enters repeated chaotic states of mind. From a dynamical point of view, these can be considered "strange attractor states" – neural net configurations that are widely distributed throughout the system and that have become engrained, repeated states of dissociated and dysfunctional activation.
>
> (p. 294)

2 While Summers has focused his discussion of futurity on its anticipatory dimensions in the developmental trajectory of human agency, he has left open the question of the history of its cognitive origins. Harari (2015) begins to fill this void with his popular book, *Sapiens: A Brief History of Humankind*. The subject matter is a sweeping evolutionary history of the human experience told in a narrative filled with imaginative generalizations. As you might imagine, the preliterate part of the story is highly speculative, "all sparkling conceptual schemas and ironic apothegms" (Mann, 2015, p. 1), with an absence if individual biography. *Sapiens* is organized around three grand revolutions: the Cognitive Revolution, the Agricultural Revolution and the Scientific Revolution. The Cognitive Revolution, about 70,000 years ago, saw the emergence of fictive language among hunter-gatherer tribes of different human species. The Agricultural Revolution, approximately 12,000 years ago, witnessed the domestication of plants and animals through the efforts of the one surviving species, *homo sapiens*. The Scientific Revolution begins with a single event, Columbus' discovery of North America some 500 years ago. It should be noted that

this origin is contested as most historians believe the scientific revolution to have been gradual, over perhaps two thousand years. However, unleashed by this event was a quest for power, knowledge and domination through the forces of colonization, imperialism and capitalism.

Our concern is with the Agricultural Revolution. Harari contends that the shift to agrarian life required the sedentary farmer pay close attention to future conditions: weather patterns, seasonal intervals, planting and harvesting cycles and defending crops and livestock against drought, pestilence and hostile neighbors. The extreme stress of such a life was lessened by living with multiple families in communal settlements for protection and survival. "Peasants were worried about the future not just because they had more cause for worry, but also because they could do something about it" (Harari, 2015, p. 101). Collective measures could be taken to lessen risks. This led to far-reaching consequences, the development of large-scale political and social systems where organization and planning were vital. An obsession about the future and shaping it to one's advantage was now the *sine qua non* of existence, something our earlier nomadic hunter-gatherer forbearers could not conceptualize.

3 It is understood that developmental theories with invariant stage sequences emerge from a positivist philosophical orientation. They purport to offer empirical data to support the theory which is taken as nomothetic knowledge, a law of nature, applicable universally. The relational and intersubjective schools, on the other hand, are embedded in social constructivism where such certainty in theory construction is considered impossible. In social constructivism knowledge is idiographic or finding patterns applicable only to specific settings and so is contextualized. It has been thought that, consequently, there are no developmental models that cohere to a social constructivism philosophical orientation. Benjamin's developmental trajectory of recognition may be the lone exception. For background see Palombo, Bendicsen and Koch (2009, pp. xi–xii, 355–356; and Bendicsen, 2013).

4 I concluded my book, *The Transformational Self: Attachment and the End of the Adolescent Phase* (2013), with a set of speculations on the nature of the Transformational Self (TS).

> One of my reviewers, my daughter-in-law, Elizabeth, asked a vital question, "If one can acquire a transformational self, can one lose it?" Also a colleague, Rita Sussman inquired, "suppose one never acquires a transformational self?" Well ... those must be subjects for a second book.
>
> (p. 196)

After the book's publication other colleagues weighed in. Sheldon Isenberg asked if the TS might appear later in life, perhaps at retirement (Personal communication, dated April 21, 2013). Barbara Alexander wondered if the TS might manifest at significant developmental junctures along the life span (in an interview with *On Good Authority*, dated July 19, 2013).

In thinking about an expanded developmental role for the TS, I saw a coming-of-age movie, *The Way, Way Back* (2013). In the lead role is withdrawn, socially awkward Duncan (played by Liam James), a 14-year-old pubescent struggling to adapt to his parents' divorce. On an extended holiday he manages to land a summer job at a water park performing maintenance tasks. During his employment he wins the "employee of the month" award; his associates give him the appellation "Pop N' Lock" for his abilities to manage locker duties. The self-referencing metaphor is not self-generated, but rather is a group designation which Duncan appears to enjoy, to own and to invest with narcissistic energy. It seems to organize him, give him status and fill him with confidence. It seems to pull him forward into other life-style possibilities and self-identifications. This self-referencing metaphor, however, is socially acquired and may be considered a precursor TS, but it will not attain the full potential of the TS and its self-reshaping power until harnessed with other developmental and maturational forces in late adolescence.

Another movie, *Tiny Furniture* (2010), allows us to reflect on the issue of not having acquired a TS. Aura, played by Lena Dunham, has recently graduated from a Midwestern college with a degree in film theory she finds worthless, as she struggles unsuccessfully to find work. She returns home to her photographer mother and competitive younger sister, who is about to graduate high school and is selecting a college. It is a hectic time for the family who appears to have little need for Aura's help. Aura reluctantly searches for satisfying work and meaningful relationships. Finding neither, she falls deeper into her personal state of anomie and begins a journey into risky and dangerous social behavior. The film ends with Aura desperately searching for stimulation and meaning in her life. The absence of a self-referencing metaphor leaves her adrift and aimless with limited prospects for acquiring a self-regulating metaphor to find and channel her potential (Rotten Tomatoes movie review, November 12, 2010).

5 One may find a self-referencing metaphor in the Biblical account of the twelve-year-old Jesus of Nazareth over 2,000 years ago. The only account of this episode in Jesus' life is found in the Gospel of Luke (Craveri, 1967, p. 61). Luke was a healer who, speaking through oral tradition, is credited with the only chronological account of the life of Jesus. From the Gospel of Luke, 2:41–52:

> Now his parents went to Jerusalem every year at the feast of the passover. And when he was twelve years old, they went up according to custom; and when the feast was ended, as they were returning, the boy Jesus stayed behind in Jerusalem. His parents did not know it, but supposing him to be in the company they went a day's journey, and they sought him among their kinsfolk and acquaintances; and when they did not find him, they returned to Jerusalem, seeking him. After three days they found him in the temple, sitting among the teachers, listening to them and asking them questions; and all who heard him were amazed at his understanding and his answers. And when they saw him they were astonished; and his mother said to him, "Son, why have you treated us so? Behold, your father and I have been looking for you anxiously." And he said to them, "How is it that you sought me? Did you not know that I must be in my Father's house?" And they did not understand the saying which he spoke to them. And he went down with them and came to Nazareth, and he was obedient to them; and his mother kept all these things in her heart.
>
> "And Jesus increased in wisdom and in stature, and in favor with God and man"
>   (Holy Bible. Revised Standard Version. New Testament, 1966, p. 69).

> It is understood that the metaphor "my Father's house" refers to the Messianic mission. "For Jesus of Nazareth is not claiming Joseph, the carpenter and son of Jacob and the man standing helplessly at Mary's side in the Temple courts, as his father. Jesus is instead claiming that the one true God of the Jewish people is his rightful parent.
>
>   (O'Reilly and Dugard, 2013, p. 78)

6 A word is necessary to distinguish the use of the adjectives' cohesiveness from coherence when referring to the self. Joe Palombo has suggested (personal communication August 23, 2011) the following distinction. Cohesiveness is a quality that is inherent in both animate and inanimate objects and, consequently, refers to a property of that object. Selves, therefore, can have more or less of the cohesive property. From the standpoint of a subjective

experience, we can speak of ourselves feeling cohesive. Coherence applies to linguistic expression and a course of reasoning. The discourse and arguments people propound may be judged to be more or less coherent or incoherent. Consequently, a narrative may be coherent, but not cohesive, unless cohesion is used metaphorically. Coherence refers to the logical relation of parts, for example, of a narrative, that affords comprehension or recognition (Websters, 1995, p. 218). Coherence is not something people experience unless one uses the term to refer to the content of what the self is attempting to process. Rather, coherence, or lack thereof, is a judgment about what people say (Bendicsen, 2013, pp. 184–186).

6

DESIDERATA

Further thoughts on case formulations

Having considered the efficacy of a developmental algorithm, let us return to a traditional case formulation to gain a better perspective on the utility of the alternative and complementary case formulation I am proposing. This chapter reflects on the interchangeability of the seven elements of a developmental algorithm with other domains of knowledge depending upon the nature of the case. As different dynamics manifest, different constellations of elements in a developmental algorithm may be needed. How are the elements in an explanatory system chosen and organized?

> All clinicians personalize the systems that they have studied and chosen to use. Their therapy is a reflection of their personal life histories – the scripts, values, attitudes, and dispositions that form, mostly at a tacit or implicit level, the weft of that elusive fabric we call the psyche. None of us can entirely escape the conditions that have made us who we are and that inevitably get enmeshed in the treatment plan and the procedures that we use with our clients. For this reason, the therapist, as person, becomes the primary instrument of therapy. The techniques become secondary.
>
> (Wedding and Corsini, 2005, p. ix)

Concerning which elements to include, we are confronted immediately with the inherent and unavoidable subjectivity of the selection process. Eschewing an objective formula, the elements should conform to a "goodness of fit" test in which movement toward explanatory comprehensiveness and coherence of thought guide the clinician.

From a traditional psychodynamic/psychoanalytic self-psychological perspective

As I consider various outlines, for the sake of brevity and uniformity, I will use the four-part Psychodynamic Formulation Model outlined by Perry, Cooper and Michels (1987). Based on my clinical experience I have modified it slightly. Part one – summarizing statement includes basic patient information, the reason(s) for referral and patient strengths. Part two – description of nondynamic factors references such nondynamic dimensions as "genetic predisposition, mental retardation, social deprivation, overwhelming trauma, and drugs and any physical illness affecting the brain" (p. 544). Note that while the origins of nondynamic factors are nonpsychological, psychological issues are most certainly intimately attached to the meaning these factors have for the patient. In part three – psychodynamic explanation, an interpretation, and integration of available data is offered using one or more overlapping theoretical frameworks. A degree of speculation in the explanation is expected and modification of the basic hypothesis is permissible with additional data. Relying on both inductive and deductive reasoning, pervasive, long-term issues and themes are identified. Motives for cognitive and behavioral patterns are pondered using various models of mental functioning. I do not want to limit my explanation to "central conflicts" and so will eliminate that aspect and consider all dynamics. Part four – predicting responses to the therapeutic situation aims at a particular aspect of prognosis, the meaning and use that the patient will make of treatment. This involves probable transference manifestations and resistances that will guide therapeutic interventions.

The "Psychodynamic Formulation" part will be the locus of expanding new explanatory knowledge. Let us begin with a self-psychological formulation as this theoretical model is a vital part of a developmental algorithm. At this point a caveat is necessary. This case is not presented as an example of orthodox self psychology formulation; rather this case is informed by self-psychological conceptualizations, in particular, the tripolar self, the selfobject (experience), the selfobject milieu, transmuting internalization, the transformations of narcissism and the contemporary understanding of human needs leading to reconfigured transferences, all more compatible with postmodern, social constructivist philosophy.

Part one – summarizing statement

Myles is a 21-year-old, high school-educated, single, retail clerk with a history of parental divorce at age 13. This trauma was preceded by the emergence of visual disturbances at age 11–12 leading to the diagnosis of schizoaffective disorder, bipolar type, by age 19. The struggle to stay on the developmental track and maintain a resilient posture in the face of massive stress has been the subject of our work. Strengths are 1) his belief in the efficacy of the selfobject milieu, 2) his anchors into the future, pulling him forward, including his wish for a promotion at work and his wish to return to college, 3) his determination to succeed and 4) his ability to narrate his life. Up until the parents' divorce, Myles had experienced, developmentally, an "average expectable environment" in the context of a secure attachment. Where this has been less than optimal, such as in the parents' divorce and once-distant relationship with his father, substantial improvement has been seen in overcoming this trauma. The full impact of the second trauma, his "condition," has been somewhat blunted by his use of the extension of the original secure attachment, his network of supporting family members constituting a therapeutic selfobject milieu.

Part two – description of nondynamic factors

Because the overarching brain disease process takes center stage, let me outline Myles' functioning against the schizophrenic syndrome and then proceed to psychodynamics. The five major symptoms of schizophrenia are delusions, hallucinations, disorganized speech (e.g., frequent derailment or incoherence), grossly disorganized or catatonic behavior and negative symptoms, such as diminished emotional expression or avolition (DSM-5, 2013, p. 99). Of the five major symptom groups only hallucinations (atypical visual distortion and brief, episodic auditory experiences involving hearing his name called) are present along with infrequent bipolar manic thoughts and depression. Myles' job, along with psychotropic medications, exposure to a functional selfobject milieu (with resonating echoes of his secure attachment) and twice weekly psychotherapy, serve to immunize him against more serious psychotic process à la Sullivan's chum effect (Sullivan in Palombo, Bendicsen and Koch, 2009, p. 235). Behavioral patterns and cognition remain organized and coherent enough to keep him employed at an unusually high level of proficiency. The cognitive impairment, usually a prominent feature of schizophrenia, has, to date, manifested in no significant way. Speech coherence has never been compromised.

Part three – psychodynamic explanation

The significant overarching stressor in Myles' life is his "condition." Anxiety, as experienced with auditory hallucinations in which his name is called from behind, is understood as a potential threat to the integrity of the self. Severe stress, such as the intensification of visual static or the deepening of his depression into suicidal proportions, is furthermore understood as disintegration anxiety with the loss of self boundaries, severely compromised self-esteem and self-confidence, and the terrifying potential for the complete dissolution of the self.

Earlier ambitions and ideals that shape the self take on young adult form having been filtered through the lens of late adolescent reality. How can we think of the passage into maturity, only now associated with young adulthood? "The hallmark of maturity is the positive self-esteem individuals feel because they are secure in the knowledge of who they are, of what they want from life, and what they feel capable of achieving" (Palombo, Bendicsen and Koch, 2009, p. 275). Beyond the biological adaptation to survival, there is the acquisition of ideals and values deemed more valuable than life itself. The perpetuation of these values transcend life itself and without which human existence has little value.

> For those who can maintain a sense of cohesion and a consolidation of their nuclear sense of self, the avenues are open for the appreciation of humor, for the possibility of creativity, for the expression of wisdom, and for the capacity for empathy for others.
>
> (Palombo, Bendicsen and Koch, 2009, p. 276)

The acquisition of these qualities and capacities proceed at an uneven pace and are acquired in varying degrees of completeness. Of course, those struggling with severe biological deficits will experience a significant imbalance in the tension between ideals and ambitions that will adversely affect the cohesiveness of the self. Myles illustrates this struggle for self-cohesion and self-regulation; it is a struggle fortunately played out in a functional selfobject milieu that can facilitate the development of compensatory mechanisms.

The spontaneous emergence of the second self-referencing metaphor, "Top dog on the floor," is a key developmental benchmark. Unlike the first self-referencing metaphor, "I keep the rhythm," which served intermediate functions, that of shaping a peer group focus and serving as a precursor to identity formation, the second metaphor is accompanied with vitality, narcissistic grandiosity and the joyful affect of ambitious achievement realized. With time and further experience, Myles' grandiosity will become reinforced with the promotion and tempered or attenuated with the reality of the need to fit into a complex team effort. In other words, the grandiose–exhibitionistic self, with adequate mirroring from peers, supervisors and therapist, will become transformed into healthy ambitions. The idealized parental imago will undergo further internalization and become strengthened into already well-formed ideals and values. As he contributes to the aggrandizement of the team accomplishments, he will also enhance the self-state's cohesion and resilience. As other self-referencing metaphors appear such as the "Top dog in the new store" and "I feel like Chapman" (the ace closer for the Chicago Cubs in the 2016 baseball World Series), the metaphors evolve from self-referencing to self-regulating, channeling potentials and lending focus and consistency to life-style choices and directions.

Part four – predicting responses to the therapeutic situation

With respect to the nature of the transference, let me approach this subject from the three sets of selfobject needs. The earliest developmental need is for mirroring. Myles has a keen sensitivity for and receptivity to confirming and validating responses from the therapist. These responses sustain him and bolster action in his daily life. The next developmental need is for idealization. This need is palpable. Myles absorbs the therapist's regulatory presence and calming influence. The therapist exists in a meaningful way outside of the session; Myles will usually call me when in crisis. The need to merge with and be like another, the twinship need, is present, but is not felt much by this therapist. It is a need better met by interaction with his uncle and significant others. All needs are present and all are being responded to in resonating fashion.

As Myles experiences his needs getting responded to he will come to feel more and more attached to those individuals, in particular his work supervisors, his AA sponsor and his uncle, who constitute the core of his selfobject milieu. Resistances to treatment will not be formidable, as long as Myles experiences his anxiety as manageable. Should anxiety intensify, his potential for regression (or fragmentation) will increase as self-cohesion is threatened. Health for Myles is understood as the development of a functional sense of agency that promotes self-cohesion and maintains his sense of well-being. Therapeutic interventions will need to be supportive and sensitive to Myles' extraordinary vulnerability. Last, the appearance of the self-referencing metaphor signals the instantiation (activation) of the Transformational Self and the beginning of the transition into young adulthood.

From complementary orientations

In trying to understand Myles' history, developmental background, symptom array, precipitants for crises and multiple diagnoses, it occurred to me that no traditional theoretical concept or set of concepts could be utilized to organize an overall approach.[1] However, approaching Myles' complex diagnoses, there might be a concept from the emerging knowledge domain of neurobiology that would help unify our work. If we think about Myles' difficulties as a brain dysregulation problem we might be able to sidestep customary theoretical turf issues and allow for a fresh approach. It is now well known that the brain is a use-dependent organ, highly sensitive to internal as well as environmental stimuli.

In the two brief case formulations that follow, I illustrate the ease with which a different explanatory dimension can be marshaled to complement a developmental algorithm. Of the Perry et al. formulation, parts one, two and four remain the same and so will not be repeated. Only part three will change.

Psychodynamic explanation – the social brain perspective

I believe that Myles' brain integration had been compromised and was functioning in a seriously dysregulated manner. It could be said that the autonomic nervous system had become destabilized. As the alarm bells of the sympathetic subsystem became activated in crisis/danger situations, the companion parasympathetic subsystem did not activate (in sufficient strength) when danger subsided. So the sympathetic subsystem did not deactivate; it remained on line with excess cortisol washing over the system requiring external structure to deactivate. One of the central roles of the prefrontal cortex is to rationally appraise the environmental situation and, when safe, send a signal to the autonomic system that it should reset itself. This signal was either too weak or not being sent. The modulating role of the prefrontal cortex needed to be strengthened. With this diagnosis the treatment plan shifted to facilitating multiple regulating experiences so as to restore brain circuitry integration.

Over the millennia, in the language of the social brain, Cozolino, using Porges' polyvagal theory of social engagement (1998; 2001; 2003), frames a slightly different explanation. Porges posits the sequential evolution of three separate autonomic subsystems. The first is the vegetative vagus unmyelinated system which controls bodily shutdown and immobilization and depends on parasympathetic processes. The second is the fight/flight sympathetic branch of the autonomic nervous system. The third is the social engagement system, a myelinated branch of the vagal system that exerts an inhibitory, calming influence on sympathetic arousal. This calming effect is called the vagal brake. The social engagement system is connected in a vast array of linkages with the face, mouth, inner ears and eyes, to exert a modulating force on visceral, emotional and behavioral states that support sustained social contact (Cozolino, 2006, pp. 87–89). The fine tuning of the vagal brake seems to depend on the quality of the attachment relationship in early childhood. By extension, in the therapeutic relationship the nature of patient-therapist interaction can exercise vagal brake processes to better regulate the social engagement system. In the case of Myles, it may be said that the emphasis on strengthening self-regulation processes may, in all likelihood, result in internal structure building consistent with that obtained in a secure, dyadic attachment relationship (Cozolino, 2006, pp. 146–148).

Myles reported that a coworker with a dysfunctional past said that his life was transformed when he discovered the alarm clock. He began using it consistently, acquired regular sleep patterns, was enjoying a life of less anxiety and was more purposeful and effective at work. This testimonial seemed to validate for Myles the benefits of a regulated life. In individual therapy with me, I encouraged Myles to remain conversational, to resist the powerful tendency to sleep all day due to his medications, and to keep commitments and appointments. Above all we worked on keeping a sense of hopefulness about the future. He was mature beyond his years and, with the disorder stabilizing, we could build compensatory mechanisms. He continued to visit his relatives, especially his grandmother, during his darkest moments. He found her unconditional acceptance and that of his girlfriend, Cindy, immensely reassuring, as stabilizing and dependable selfobjects. He needed to test his growing capacities against the strength of the disorder. For example, despite his body tremors, he managed to engage customers, meeting his metrics. He impressed management to enhance his reputation as the "Top dog" and consider giving him a promotion. I helped to prepare him for possible disappointment, but his initiative never flagged. I explained to Myles something of the nature of "how the mind works." Myles found universalization helpful, enabling him to feel that some of his anxiety is a normal part of human experience. It is hoped that this developmental process of promoting self-reflection in therapy may strengthen the prefrontal cortex's executive role in relation to the autonomic system and promote better reality testing, planning and enhance self-calming abilities. He wanted very much to minimize the descent into cognitive confusion, paralyzing depression and intense, directionless anxiety. The frightening descending spiral into despair and panic could now be influenced and moderated with a set of expectations about the future that Myles believed he could influence and shape. The goals of treatment are maintaining hopefulness, enhancing self-regulation through the use of calming selfobjects, and development of a realistic ego/self ideal.

Let us now move to a consideration of another complementary developmental framework, the non-linear dynamic systems perspective that offers enhanced explanatory usefulness.

From a complementary non-linear dynamic systems perspective

A dynamical systems hypothesis of schizophrenia has emerged (Loh, Rolls and Deco, 2007). In proposing models for schizophrenia, one of the difficulties is the complexity and heterogeneity of the illness. The authors have developed a novel approach that does not rely purely on biological mechanisms, but, rather, links the inconsistent symptoms in a top-down approach based on instabilities in attractor neural networks.

> Our hypothesis is based on the concept of attractor dynamics in a network of interconnected neurons that in their associatively modified synaptic connections store a set of patterns, which could be memories, perceptual representations, or thoughts (Amit, 1989; Hopfield, 1982; Rolls and Deco, 2002). The main assumption in our hypothesis is that attractor dynamics are important in cognitive processes such as short-term memory, attention, and action selection (O'Reilly, 2006; Deco and Rolls, 2005). The network may be in a state of spontaneous activity, or one set of neurons may have a high firing rate, each set representing a different memory state, normally recalled in response to a retrieval stimulus. Each of the states is an attractor in the sense that retrieval stimuli cause the network to fall into the closest attractor state, and thus to recall a complete memory in response to a partial or incomplete cue. Each attractor state can produce stable and continuing or persistent firing of the relevant neurons.
>
> The authors adopt the concept of a three dimensional energy landscape metaphor with hills and valleys (basins). "The concept of an energy landscape (Hopfield, 1982) is that each pattern has a basin of attraction, and that each is stable if the basins are far apart and also if each basin is deep, which is caused, for example, by high firing rates and strong synaptic connections between the neurons representing each pattern, which together make the attractor state resistant to distraction by a different stimulus. The spontaneous firing state, before a retrieval cue is applied, should also be stable. Noise in the network caused by statistical fluctuations in the stochastic [or random] spiking of different neurons can contribute to making the network transition from one state to another; we take this into account by performing

integrate-and-fire simulations with spiking activity, and relate this to the concept of an altered signal-to-noise ratio in schizophrenia (Winterer, Coppola, Goldberg, Egan, Jones, et al. 2004; Winterer, Musso, Beckman, Mattay, Egan, et al. 2006; Winterer, Ziller, Dorn, Frick, Mulert, et al. 2000)."
(Loh, Rolls and Deco, 2007, pp. 1–2)

The authors consider how the three main symptom clusters in schizophrenia, cognitive dysfunction, negative symptoms and positive symptoms, might be produced in a neurodynamical system (Baxter and Liddle, 1998; Liddle, 1987; Mueser and McGurk, 2004). Working memory dysfunction may be related to instabilities in the firing rates of attractor networks in the prefrontal cortex.

> The neurons are firing at a lower rate leading to shallower basins of attraction of the persistent states, and thus a difficulty in maintaining a stable short-term memory ... The shallower basins of attraction would thus result in working memory deficits, poor attention, distractibility, and problems with executive function and action selection.
> (Durstewitz, Seamans and Sejnowski, 2000; Wang, 2001; Loh, Rolls and Deco, 2007, p. 2)

Flattening of affect and a reduction in emotion may be caused by a consistent prefrontal hypo-metabolism or hypo-reduction in firing rates of neurons in the brain region associated with emotion. "We propose that these symptoms are related to decreases in firing rates in the orbitofrontal cortex and/or anterior cingulate cortex where neuronal firing rates and activations in functional MRI investigations are correlated with reward value and pleasure" (Rolls, 2005; Loh, Rolls and Deco, 2007, p. 2).

Hallucinations or delusions may be caused by instabilities in both spontaneous and consistent attractor states associated with higher activity in the temporal lobes.

> We propose that these symptoms are related to shallow basins of attraction of both spontaneous and persistent states in the temporal lobe semantic memory networks and to the statistical fluctuations caused by probabilistic spiking of neurons. This could result in activations arising spontaneously and thoughts moving too freely about the energy landscape, loosely from thought to weakly associated thought, leading to bizarre thoughts and associations that may eventually over time be associated together in semantic memory leading to false beliefs and delusions.
> (Scheuerecker et. al. 2007; Shergill et al., 2000; Loh, Rolls and Deco, 2007, p. 2)

"There are specific symptoms such as aberrant eye movements that cannot be accounted for by this general scheme" (Loh, Rolls and Deco, 2007, p. 2). Therefore, and regrettably, a potential explanation for Myles' severe optic disturbance remains unavailable through this dynamical systems hypothesis.

The application of dynamical systems to mental illnesses, such as schizophrenia, offers perhaps the most contemporary theoretical explanation to this most baffling, multifaceted brain disease.

Psychodynamic explanation – non-linear dynamic systems perspective

In a neurodynamical system a state of homeostasis is presumed to exist which is in a constant state of reconfiguration and rebalancing. In this orientation, it is hypothesized that internal and external attractor forces continuously perturb the system in ways that contribute to stabilization or destabilization. Myle's auditory hallucinations may be attributed to instabilities in both spontaneous and consistent attractor states associated with higher activity in the temporal lobes. The instability may cause weakly connected ideational associations to flow more freely about the energy landscape. An instability in the motor cortex function of the temporal lobes may account for the occasional manic flight of thoughts and activity. Myle's tendency to lose awareness of the necessity for daily social hygiene may be attributed to an executive function deficit in the prefrontal cortex. Myles' overall condition may be understood as a strange attractor, one that is highly complex and suggesting an underlying global order to the system behavior, but a local symptom unpredictability.

Over the millennia, and especially the past two hundred years, attempts to explain schizophrenia have existed *pari passu* with efforts to categorize the forms of the disease (Arieti, 1974, pp. 9–29). I have presented two of the latest attempts to explain schizophrenia: the social brain hypothesis and the non-linear dynamic systems hypothesis. I suggest that these two explanations should complement rather than compete with each other. Rather than struggle to integrate these approaches, let us consider a pluralistic perspective. Goldberg (2007) has given pluralism some thought.

> Pluralism is a philosophical doctrine that says there is no one principle that underlies all forms of thought. Thus, much like the overdeterminism of behavior that is familiar to all psychoanalysts, there need not be a single explanation to encompass all theories or techniques of therapeutic action. Therefore, an improvement in a patient's well-being as a result of psychoanalysis can be explained as a byproduct, a reaction to the warmth of the analyst's personality, a developmental achievement, an example of the efficacy of insight, a learning of how to handle discouragement, or all other manner of explanatory devices.
>
> (Goldberg, 2007, pp. 1667–1668)

While any device can carry the weight of explanation, all devices are not of equal measure. The quest to use multiple theories in psychoanalysis is dogged by insularity not unlike the contemporary tribalism in politics. As is the case with postmodernism, "there are no overarching theories that cover everything, but one needs to see what works best under what situations" (p. 1670). How is it decided what works best for each patient? "Pluralism answers this question by alerting us to the possibility that psychoanalysis must be seen as an evolving set of concepts with varying applicability. What psychoanalysis needs is genuine scientific pluralism" (p. 1673).

In the absence of a cure for the baffling brain disease of schizophrenia, I suggest there is, nevertheless, a general consensus on a treatment protocol which is applied to the case of Myles in Chapter Seven.

Note

1 The history of theoretical speculation over the etiology of schizophrenia offers us an example of how multiple theories compete rather than complement each other for explanatory power. Some of these more compelling theories possess an adhesive quality that makes it more difficult to discard them in favor of more modern, empirically oriented theories. The modern formulation of schizophrenia can be said to begin with Freud who struggled with applying the psychodynamics of the conflict model (Freud, 1924). Was schizophrenia a response to conflict with others leading to intense frustration and regression to the autoerotic stage of objects relations development? Or was schizophrenia a structural psychic defect originating in an unbearable discordance between the ego and external world explained by a withdrawal of object cathexis and a reinvestment in the ego. Sullivan (1962) advanced the belief that the etiology of schizophrenia could be found in interpersonal difficulties due to faulty mothering. This mother generated great anxiety, was unresponsive to her child's basic needs and created a self-experience that was dissociated. Damage to self-esteem was severe, but the schizophrenic patient was still left with some capacity for interpersonal relatedness. And, unlike Freud, Sullivan believed limited transferences could be formed. Federn (1952) emphasized the withdrawal of ego boundary hypothesis, suggesting that schizophrenic patients could not discern inside from outside. Weakland (1960) teamed up with family systems researchers and communication experts and formulated the notion that schizophrenia originated in confusing, inconsistent family pathology. The mother/family sent the child double-bind messages so that whatever decisions the child made, they were the wrong ones. Eventually, theories of schizophrenia incorporated biological elements, leaving us with the contemporary explanatory hypothesis that schizophrenia is "probably" a brain-based disease that develops in "an interaction among genetic vulnerability, environmental attributes and individual traits" (Gabbard, 2005, pp. 183–187; see also *The Etiology of Schizophrenia*, 1960, edited by D. D. Jackson).

7
TREATMENT EFFICACY

The treatment of schizophrenia

What do practice guidelines recommend regarding the treatment of schizophrenia spectrum disorders? According to the *Harvard Mental Health Letter* (*HMHL*) (November, 2008; 2010) there is consensus that a two-pronged approach including 1) the administration of psychotropic medications plus, 2) the application of a set of psychosocial interventions, offer the optimal chance of success. With respect to front-line medications, second-generation atypical antipsychotics such as clozapine (Clozaril), olanzapine (Zyprexa), quetiapine (Seroquel), risperidone (Risperdal), sertindole (Serlect), ziprasidone (Geodon), aripiprazole (Abilify) iloperidone (Fanapt) and paliperidone (Invega) are considered helpful in managing positive, and negative symptoms, and short-term management of behavior disturbances.[1] Clozaril is the most effective second-generation antipsychotic, but its use in first episode schizophrenia is not recommended because of its substantial side-effect profile including weight gain, diabetes, sedation and many other medical problems (*HMHL*, 2008; *HMHL*, 2010; Dziegielewski, 2006, pp. 173–175). While many exceptions exist as to efficacy of first-generation versus second-generation antipsychotic medications (*APA Practice Guidelines*, 2004 and *Psychiatry Online Guideline Watch*, 2009) and individual tolerance, all studies agree that medication adherence is essential to the opportunity for successful treatment (*HMHL*, 2008).

Regrettably, extrapyramidal symptoms, in particular body tremors, are a common, usually irreversible, side-effect of antipsychotic medications. Rhythmic and repetitive leg and arm movements, suggestive of tardive dyskinesia, are frequent (Dziegielewski, 2006, pp. 169–171). Myles has such movements. He is quite self-conscious of these obvious motions and has developed ways of sitting and standing that limit their appearance in social and work settings.

Regular checkups, screenings and treatment for medical comorbid conditions is an imperative.

> Life expectancy for both males and females has been increasing over the past several decades to an average of 71 years. However, the life expectancy among individuals with schizophrenia in the United States is 61 years – a 20% reduction. Patients with schizophrenia are known to be at increased risk of several comorbid conditions, such as type 2 diabetes mellitus (T2DM), coronary heart disease, and digestive and liver disorders, compared with healthy people. This risk may be heightened by several factors, including sedentary life style, a high rate of cigarette use, poor self-management skills, homelessness, and poor diet.
> (Kahn et al., March, 2016, p. 31)

Also, comorbidity must take into account increased risk for obesity, chronic obstructive pulmonary disease, hepatitis C, aids and cancer (Kahn et al., March, 2016, pp. 30–40). To this list of medical risks must be added the need for good oral hygiene to mitigate edentulism, tooth loss (Raza, Hirapara and Hussain, April, 2016).

Concerning psychosocial treatment, the "strategies are meant to support a patient's ability to learn to live with schizophrenia, and often involve long commitments of time" (*HMHL*, 2010, p. 5; *APA Practice Guidelines*, 2004 and *Psychiatry Online Guideline Watch*, 2009). The interventions include:

1) Assertive community treatment involving a multidisciplinary team, low patient-to-staff ratios and frequent patient contact;
2) Supported employment, training and rapid job placement;
3) Skills training centered on social and interpersonal skill building using positive reinforcement, feedback and frequent practice;
4) Cognitive behavioral therapy in individual and/or group modalities especially for those who continue to experience psychotic symptoms;
5) Token economy interventions focusing on personal hygiene and other behaviors that facilitate adjustment to real-world conditions;
6) Family psychoeducational services that facilitate ongoing family interaction and collaboration among family members;
7) Alcohol and substance abuse interventions that center on motivational enhancement and behavior therapy;
8) Peer support and peer-delivered services recognize the value of utilizing the experience of individuals who have severe mental illness in the treatment process; and
9) Weight management to counteract one of the most frequent side-effects of antipsychotic medications.

The most successful psychosocial interventions involve intensive and consistent (more than two years) patient interaction.

Let me expand on number 6, the section on family psychosocial services. Marley (1994/2004) has developed a comprehensive outline of family involvements in the treatment of schizophrenia. While many of these approaches seem academic and too difficult to implement in orthodox form, one stands out as being imminently practical, that of the Psychoeducational Model. We learn that this model was developed directly from work with families struggling with schizophrenia and so has a proven pedigree. It is a strengths-based, collaborative approach emphasizing structured family meetings and planned roles for each family member. As tasks are assigned, the executive capacities of each family member are realistically appraised for successful implementation (Minuchin, 1974). Elements in the approach include dissemination of accurate information, readjusting expectations, emphasizing collaboration within and without the family system, building social supports and developing flexibility. Treatment dimensions include providing a survival skills workshop, practicing reentry back into the family environment post hospitalization, immersion into social and vocational rehabilitation and meeting established goals in the context of recognizing and celebrating achievements (Marley, 1994/2004, pp. 107–113).

From these practice guidelines it is clear that Myles, while afflicted with a severe mental illness, remains a high-functioning individual who is strongly motivated to improve. His family also is high functioning. His treatment plan contains essential psychoeducational elements and is in alignment with best practice guidelines. Also, the elements of the plan are interacting in such a way as to capture and harness therapeutic synergies. He receives an annual physical exam from his family doctor and periodic screenings to manage the psychoactive medications.

In conclusion I want to emphasis two points. First, a knowledge of regulatory systems informs my therapeutic technique by privileging the interaction of multiple methods and collaborators, over a strictly hermeneutic, intrapsychic perspective, so as to capture system synergies. In order to stabilize and regulate the organism, the interaction of a host of biopsychosocial-spiritual variables by an "implicit" organizer is necessary. The implicit organizer is one that emerges as a result of consensus on the supportive treatment protocol. In this protocol the selfobject milieu system functions with a unitary impact, the elements enhancing each other's effects. This offers a useful metaphor for the need to keep a variety of system variables in alignment to achieve optimal therapeutic gain. The objective of capturing and employing functional synergies, as the central therapeutic tool, is considered absolutely necessary in confronting leviathan disturbances such as that described with Myles.

Second, by valuing multiple approaches and developing therapeutic alliances with a range of collaborators, the long-term trajectory of the disturbance is addressed. Helping forces will be needed throughout Myles' vulnerable life. Regulation theory, as a treatment organizer, keeps the treatment balanced and focused on the whole, ongoing, contextual life of the client.

Note

1 The controversy between the benefits against the risks of short-term versus long-term psychotropic medication management continues. In Whitaker's *Anatomy of an Epidemic: Magic Bullets, Psychiatric Drugs, and the Astonishing Rise of Mental Illness in America*, an alternative viewpoint is expressed with regard to the efficacy of psychotropic medications. Whitaker maintains that the explosion in mental illness in America over the past fifty years is due in large part to the over-prescription of psychotropic medications. His extensive review of the research literature and interviews with the researchers strongly suggests that the best use of psychotropic drugs is for short-term stabilization of the patient. Psychotropic "drugs, rather than fix chemical imbalances in the brain, perturb the normal functioning of neurotransmitter pathways leading to worsening outcomes" (Whitaker, 2010, p. 333). It is my belief that Myles benefits substantially from the psychotropic mixture he is presently taking. In the short term he is realizing far more stabilization and symptom reduction than he would without the medications. The long-term effects of this potent mixture is potentially troublesome and will, therefore, require careful monitoring and ongoing consultation with the prescribing psychiatrist.

8
SOME THOUGHTS ON THE ADOLESCENT PASSAGE, EMERGING ADULTS AND MILLENNIALS

While the manifold types of transitions into adulthood fill a spectrum, it seems that contemporary cultures are evolving two essential ways by which adolescents transition to adults. In the first way, I am referring specifically to adolescent coming-of-age rituals which occur either at the beginning or the ending of adolescence. These are the clear social initiation rites of passage which are found throughout the world. In the traditional Jewish religion the bat mitzvah for girls at 12 and the bar mitzvah for boys at 13 signifies becoming a full-fledged member of the Jewish community. The b'nai mitzvah coincides roughly with puberty. The initiate, having successfully finished a period of study, is now accountable for their actions (Kaplan, 2007, pp. 164–167). At the Dongmyeong Girls Middle School in South Korea, a coming-of-age ceremony is combined with graduation, marking the passage into adulthood (Jones, 2015). In Latin America, primarily Mexico, the Quinceanera celebrates a girl's fifteenth birthday, marking the passage from girlhood to womanhood. It is both a religious and social event combining aspects of ancient Mayan culture and Roman Catholicism (Kalman, 2008; Encyclopedia Britannica Online).

In Occidental culture the coming-of-age theme emerges from the *Bildungsroman* genre of German literature. While there are numerous examples of such literature, the archetype appears to be *Wilhelm Meister's Apprenticeship* by Johann Wolfgang Goethe written in 1795–1796 and published in 1821. The term *bildungsroman* deals with the maturation process and means "novel of education" or "novel of formation." The theme centers on the psychological and moral growth of the protagonist of either gender from youth to adulthood in which character change is extremely important (Encyclopedia Britannica Online, December 10, 2015). Each generation seems to produce powerful coming-of-age stories of protagonists from Telemachus through Stephen Dedalus, Nanook, Holden Caulfield to Jean Louise "Scout" Finch. Finishing high school, entering a trade or graduating from

FIGURE 8.1 Dustin cartoons illustrating the prolongation of the adolescent phase.
Reproduced with kind permission of King Features Syndicate

college and securing the first full-time job are common secular rites of passage dimensions of the coming-of-age theme. In these examples a cultural imperative is engaged, a struggle/conflict ensues with formative challenges addressed, culminating in a positive ending, sweeping the adolescent along into adult expectations and behaviors.

The second way seems to involve deferring major decisions about one's life and, in so doing, elongating the adolescent phase. A significant cause of this delay is obviously the monumental economic recession of 2008. For those in this involuntary circumstance, life is not yet goal-driven and focused. For these adolescents Arnett (2004) proposes a new developmental stage he calls "emerging adulthood."

> The emerging adulthood stage is interposed between adolescence and adulthood as a way to explain the delay in making lifelong commitments traditionally characteristic of adulthood. A psychological profile accompanies those in emerging adulthood consisting of identity exploration, instability, self-focus, feeling in between and a reflective element, a kind of poetic musing he calls a 'sense of possibilities' (Arnett, 2004, pp. 7–17). What can be said of this clustering of dynamics? I believe that the attempt to create a new developmental phase, emerging adulthood, is taking too great a conceptual

leap. The cluster of dynamics seems consistent with those associated with adolescence and not sufficiently distinctive to justify a new phase.

Erikson's (1958) concept of **'the psychosocial moratorium'** (pp. 43; 100–104) contains adequate conceptual power to clarify the prolongation of the adolescent phase. In Erikson's words, a moratorium is "a span of time after they have ceased being children, but before their deeds and works count toward a future identity." A variety of moratoria is possible. The adolescent may not know that they are marking time before they come to their crossroad, which they often do in their late twenties, belated just because they gave their all to the temporary subject of devotion, as was the case for Martin Luther (p. 43). Of course, adolescents may choose to suspend their development for many reasons, some in the service of healthy "identity" consolidation, others in the cause of more pathological "identity confusion" (Erikson, 1950, pp. 261–263; (Bendicsen, 2013, pp. 193–194). For an interesting discussion of the argument as to whether Arnett's emerging adulthood concept is a stage or a process see Harter (2012, pp. 155–157).

Let us look at late adolescents from a sociological perspective. The adolescents we study are part of a cohort, a target population that now seems to have acquired an adhesive name: the Millennials. Also known as Generation Y or the Me Generation, the Millennials, some 80 million strong, have emerged as the largest of the postwar generations. The Millennials were born, give or take, between 1980 to 2000 and are now between 15 and 35 years of age. Larger than the preceding generation, Generation X, those born between 1961 and 1980, and larger than the Baby Boomers, those born between 1943 and 1960, the Millennials are a consuming and purchasing bonanza for advertisers and marketers who are eager to accommodate novel tastes and the whole range of nontraditional and alternate life-style experiences.

Millennials seem to prefer inexpensive hostels and "lifestyle hotels" over conventional hotels (Mayerowitz, April 17, 2015; Herbling, May 23, 2015; Janssen, January 26, 2016). Las Vegas hotels are trying to appeal to millennials by installing an e-sports venue featuring competitive video game tournaments (Cano, February, 28, 2017). Millennials seem to prefer fast, casual dining involving unconventional mixes of snack foods (Cancino, May 17, 2015; Herbling, May 20, 2015). Chuck E. Cheeze, the kiddie pizza chain, has a solution to its sales slump: winning over millennial moms (Giammona, October, 5, 2015). McDonald's is unveiling a new packaging format to attract millennials (Rosenthal, January 8, 2016). About 35% of millennials live at home with parents. They tend to marry later and carry more college debt. Many are choosing to have children later or not at all, causing researchers to worry about a national decline in the birthrate (Cha, July 2, 2017). Millennials are avoiding the use of credit cards (Fu, March, 18, 2018). Many consider themselves underemployed or find it difficult to find and keep meaningful, full-time employment (Manchir, 2013; Keilman, November 11, 2015). Another commentary places the number of millennials who live with parents at 26% and regard that level as significant enough to be slowing down the economy because they tend not to purchase new goods (MarksJarvis, August 2, 2015). "A Pew Research Center analysis of U.S. Census Bureau data found that 36.4 percent of women between the ages of 18 and 34 lived with parents or relatives in 2014, the most since at least 1940" (Sell, November 12, 2015; Sichelman, January 31, 2016). Millennials are showing a lesser interest in riding Harley-Davidson hogs and are inclined to use ride-sharing car services or hop on a Divvy bicycle (Reed, July 20, 2017).

Is the psychological effect of millennial children returning home necessarily negative? For some, both parents and children, the reunion is enjoyable (Martin, J., July 28, 2012) For this segment the stigma of returning home is absent. And some never leave by purposely picking a nearby college so they can continue to live at home (McNamara, September 14, 2011). On the other hand, boomerang kids can interfere with parents who are trying to downsize (Martin, May 28, 2017; Martin, October 15, 2017; see also movie *Sisters*, 2015). Millennials tend to not want their parents' old furniture, a difficult prospect for parents who do not want to donate their nostalgic furniture to Goodwill (Quigley, June 17, 2016).

A recent survey revealed that 40% of millennials receive financial help whether living at home or away, married or living with a significant other (Cross, February 1, 2016; Mann, May 1, 2016). On the other hand, some see an emerging trend in which more renting millennials are assuming first time mortgages (Channick, July 1, 2015). Reaching out to millennial home buyers, Quicken Loans and Freddie Mac are teaming up to offer selected consumers a mortgage loan with a down payment of only 3% (MarksJarvis, October 21, 2015). Then again, due to student debt burdens, many millennials face a less affordable housing market than their parents, forcing them to put off home ownership (Boak, December 6, 2015; Greiwe, January 29, 2016). According to Wong, "millennials are more likely to 'choose the Starbucks and live in an attic' than 'have a nice apartment and no Starbucks'" (Wong, October 28, 2015). Many are social activists and work with the under-privileged and disadvantaged (Cancino, January 27, 2015). Millennials are avoiding purchasing food at large chain stores, will make more frequent trips to smaller merchants, and want more information on food labels (Bomkamp, November 2, 2015).

Politically, many will vote for liberal democrats (Heuvel, 2015). Having little faith in government and the other institutions they thought they could depend on, millennials are being drawn to the presidential candidacy of Bernie Sanders and socialism because of what they describe as his authenticity, his idealism and his unvarnished take on their everyday realities (Wagner, November 1, 2015; Taylor, February 22, 2016; Chapman, May 20, 2018). Many millennials believe the American dream, typically defined as the "having the opportunity to achieve a satisfying life if you are willing to work hard and play by the rules," is dead for them. "Large swaths of millennials have lost faith in the political process" and do not see themselves as change agents (Lewis, January 4, 2016). The former head of the CIA from 2006–2009, Michael Hayden, suggested the younger millennial agents may be leaking state secrets. They have cultural differences from baby boomers, viz., they tend to be less patriotic, more politically independent, religiously unaffiliated and unmarried (Berman, March 15, 2017). In a large study conducted by the Black Youth

Project at the Center for the Study of Race, Politics and Culture at the University of Chicago, black millennials, while optimistic that they can make a change through participation in politics, are more likely to be poorer and unemployed, face a greater degree of gun-related violence and discrimination (Muskal, November 5, 2015). Surprisingly, many millennials favor gun rights (Rampell, December 9, 2015).

The average age for millennial retirement is projected to be 73 due in part to lower wages, defined benefit plans giving way to defined contribution pension plans and millennials being risk-averse when it comes to investing in the stock market (Glinski, 2015). A 2015 Gallup poll found that 64% of millennials do not think the Social Security program will be able to pay benefits when they retire ("Social security chugs toward the cliff," Editorial, July 20, 2017). Stacy Rapacon (Newsweek, February 4, 2016) places the average millennial retirement age at 75. In the workplace millennials are central to a new corporate trend called reverse mentorship. "That's when millennials take older employees under their wing to teach them how most corporate revenue problems can be solved with a few social-media tricks, and why you shouldn't ever leave voice mails for anyone" (Schrobsdorff, December 14, 2015). Millennials are technologically savvy, having come of age during the information age and the domination of computers and wireless communication in daily life. Many avoid owning personal automobiles, preferring public transportation (Moore, 2015). Nevertheless, millennials are being targeted as a unique cohort, receptive to buying non-popular, unsuccessful, even discontinued automobile brands (Williams, October 24, 2015; Phelan, August 20, 2017).

Millennial parents in the digital age "are weighing whether to snap up a domain name for their baby, post photos that family members can click through or stow for later or create a slot for their child on social media sites like Facebook or Twitter" (Bowen, November 1, 2015). Millennials as parents are marked by optimism, have faith in progress, equality and Google and build vast archives of selfies. Pressure to be competent, even great parents is fierce. They believe if everything is done correctly then everyone will be safe. Qualities they want their children to acquire are to be open-minded, empathetic, questioning and unique. Preferring to cohabit with partners rather than marry, millennials are shifting from the helicoptering parenting style of their parents to that of drone parenting, a style in which "the parents still hover, but they're following and responding to their kids more than directing and scheduling them" (Steinmetz, 2015, p. 41). Many millennials, now with children and families, are moving out of the city and back to the suburbs (Clark and Greenfield, October 13, 2017). Searching for a more meaningful life, some Millennials are looking to give their life to something such as the priesthood or a religious calling, even though those who identify with the Catholic Church has fallen from 22% to 16% since 2000 (Hennessy-Fiske, September 27, 2015).

On the other hand, some change jobs frequently, feeling little loyalty to employers. In the work place, earlier generations find that many millennials are self-absorbed, self-promoting and feel a narcissistic sense of entitlement. In addition, they lack focus and are preoccupied with technology. But, with about one-fourth of leadership positions now held by millennials, a study by Future Workplace found that many millennials embrace traditional values, such as the ability to build relationships, the capacity to work well and build a great team and the development of good communication skills (Huppke, November 16, 2015). Some millennials have become successful inventors and wealthy entrepreneurs (Channick, October 7, 2017). With emphasis on social justice, will millennials be able to end sexual harassment in the workplace (Weber, October 13, 2017)? However, some millennials are as prejudiced as those in earlier generations (Page, 2015). In a survey from the Pew Research Center, millennials are more "likely to give themselves low rankings in categories such as patriotism, responsibility, willingness to sacrifice, religiousness, morality, self-reliance, compassion and political activism. Fully 59% say 'self-absorbed' is an apt description of their bunch" (Kunkle, September, 6, 2015). Another observer writes that a more rounded picture of millennials reveal a generation as diverse as previous ones, one with optimism and potential (Stein, 2013).

In summary, Sasse posits this observation: "Although it is not universally fair, millennials have acquired a collective reputation as needy, undisciplined, coddled, presumptuous, and lacking much of a filter between their public personas and their inner lives." In the workplace "managers struggle with their young employees' sense of entitlement, a tendency to overshare on social media, and frankness verging on insubordination" (Sasse, 2017, p. 125). Correcting Sasse's lack of emphasis on millennials' strengths, Stevens writes,

> Millennials volunteer their time in higher percentages than previous generations, are on track to become the most educated generation in American history and are the most open minded demographic when it comes to LGBTQ rights and interracial marriage, according to Pew Research Center statistics.
> (Stevens, May 28, 2017; see also Wilhelm, May 26, 2017)

From a studied perspective, the significance of the millennials for our exposition is that this cohort is measurably contributing to the continuation of the trend of the prolongation of the adolescent phase, a trend first noted and recommended as desirable by J. J. Rousseau (1762) and followed by G. S. Hall (1904; Kaplan, 1984 in Bendicsen, 2013, pp. 3–4, 40–41). Knapton (January 19, 2018) writes that social scientists in Great Britain, Australia and New Zealand are agreed that the age span of adolescence should be adjusted from 10–18 up to 10–24. The ages 10–24 take into account that the brain does not mature until about age 25, that wisdom teeth do not come out until after 25, and that the average man now marries for the first time at age 32.5 and the average woman for the first time at age 30.6. The massive college debt load of millennials, their struggle finding and keeping full employment and the financial need for many to continue to live at home or to return home, speaks to the cultural impact on shaping individual internal psychological forces. These forces constitute powerful determinants in the timing of the transition into young adulthood. Conventional wisdom has it that the multiple pathways and timetables to young adulthood privilege no particular method, especially for those who do not experience a rite of passage.

9
SYNOPSIS

This monograph continues and expands the body of work I laid out in the *Transformational Self: Attachment and the End of the Adolescent Phase* (2013). In this work I began by declaring that new criteria need to be formulated to account for the varieties of intrapsychic and interpersonal transformational developmental processes encountered in the transition from late adolescence to young adulthood. Traditional ego psychology explanatory frameworks such as Mahler's separation-individuation (1975) concept and the developmental recapitulation tasks of Blos (1962) and Colarusso (1992) are no longer adequate. (These traditional tasks are reviewed in *The Transformational Self*, 2013, pp. 47–51, in the context of the ending of adolescence.) I intend to shift the discussion from the static review of developmental tasks to the dynamic analysis of intrapsychic and interactional processes. My working hypothesis is: An integration of neurobiological research findings, non-linear dynamic systems theory, embodied metaphor theory, contemporary psychoanalytic developmental theory, attachment theory, self psychology with intersubjectivity and relational theory, work together to transform the sense of self and bring closure to, and movement through, the adolescent phase. I have decided to label this synergistic process regulation theory. The process of transformation is understood through a systems approach combined with an intrapsychic perspective in which the appearance of idiosyncratic personalized metaphors contribute to the adolescent's enhanced self representation which shape a focus for further development.

I introduced the concept of the Transformational Self, a phase-specific dimension of the neural self, and demonstrated the enhanced explanatory power that it offered as one attempts to examine the sometimes dramatic shifting self-states accompanying the metamorphosis from adolescence into young adulthood. A necessary precondition for the emergence of the Transformational Self is the maturation of the prefrontal cortex and its enhanced neural interconnectivity. With this biological achievement, executive functioning, a strengthened ego/self capacity, can arrive at a mature level of external stabilization and internal, intrapsychic structuralization. Conceptualized in self-referencing metaphor and expressed and reinforced through long-term potentiation (repeated firing patterns of synchronous neural assemblies), the late adolescent reconfigured self-state becomes a true developmental potentiality evidenced by the use of different self (and other) representations. In other words, self-referencing metaphor becomes the pathway to personal metamorphosis. With even a minor self-referencing metaphoric input an identity synergy is created through repeated firing of specific neural networks which can manifest in a disproportional outcome – the Transformational Self.

Empirical data gathering in the clinical office setting set in the context of a single case research design methodology shapes interpretations understood as hermeneutic inferences regarding the possible significance of the unique self-referencing metaphoric phenomena. This book then is an exercise in inductive reasoning where data from a few samples is generalized to the overall cohort of late adolescents. The psychotherapies of two mid-adolescent girls illustrated the application of the Transformational Self concept.

I proposed a two-pronged effort. First, I created a new, unitary framework of interdisciplinary processes that I label regulation theory that accounts for the structural and functional aspects of transitioning from late adolescents to young adulthood. Second, I applied regulation theory to two psychotherapy cases to demonstrate its operational usefulness from both diagnostic and treatment perspectives and, in so doing, illustrate its enhanced explanatory power for describing dynamic processes.

In this monograph, *Psychoanalysis, Neuroscience and Adolescent Development: Non-Linear Perspectives on the Regulation of the Self*, I continue and enlarge the conceptualizations I presented in the *Transformational Self*. Specifically, in this work, I expand on two themes first articulated in *The Transformational Self*: 1) I clarify the nature of regulation processes, placing them in a developmental context, and 2) I demonstrate the theoretical usefulness of using a set of interlocking theories to form a developmental algorithm, which operationalizes regulation theory, and illustrate its therapeutic efficacy through a case example. Let me briefly recapture my argument as it unfolds in the eight parts.

In Chapter One, I began by presenting a compressed review of over one hundred years of psychoanalytic theorizing to illustrate that:

1) the continued fracturing of psychoanalytic theorizing is a lasting artifact of Freud's secret committee process that created dogma insiders and outsiders;
2) the Balkanization of psychoanalytic theory has contributed to the marginalization of psychoanalysis;
3) there is no consensus on which psychoanalytic theory occupies the contemporary conceptual high ground;
4) the restoration of psychoanalysis to a position of conceptual usefulness will lie in the role it contributes to a multidisciplinary effort. A central part of that

effort must include neuroscience and a new contemporary metapsychology. The remarkable advances in neuroscience research offer a potential fulfillment of the promise held out by Freud in his *Project for a Scientific Psychology* (1895). It is clear that a vigorous intellectual and scientific effort is being made to join psychoanalysis with neuroscience. Psychoanalysis is oriented toward explicating the humanistic, hermeneutic, and subjective dimensions of the human experience. On the other hand, neuroscience is focused on research outcome replicability, empirical data collection and objectification of that same human experience. The combining of these two spheres of expanding knowledge is yielding a fruitful collaboration in the emerging domain of neuropsychoanalysis, a domain that might lead to the creation of a new explanatory paradigm (Palombo, Bendicsen, Koch, 2009, pp. 358–362; McGowan, 2014).

In Chapter Two, I shift the discussion to the new paradigm – that of regulation theory and its inexorable linkage to neuroscience. Regulation theory is actually a composite of many theories. It represents a compatible set of interlocking theories in tight theoretical alignment. I demonstrate that this alignment enables us to array disparate data in such a fashion that an opportunity is secured to provide an overarching degree of coherent explanatory power.

The modern view of the brain is that it operates as a complex dynamic system. Its elements coexist in a state of mutual co-regulation. The functionality of the system is described in the context of the Social Brain. The Social Brain conceptualization is the successor to the Triune Brain which is grounded in evolutionary and anatomical/structural dimensions. In contrast, the Social Brain describes the functionality of its regulatory subsystems and so comports well with our new regulation paradigm.

Following Siegel (1999, p. 274) I attempt to bring some measure of order to the multiple biological and cultural regulatory models by organizing five of them into an epigenetic developmental sequence. I begin with Hofer's "hidden regulators" that organize neonatal homeostasis. Hofer's framework is followed by Shore's (1994) groundbreaking hypothesis that "the primary function of attachment is that of regulating the child's affect states." Next, a necessary condition for successful school functioning is the attainment of emotional self-regulation. As emotional life evolves and discriminates, it serves as a vital organizer, integrator and regulator of meaning-making, motivational direction and shapes mutual subjectivity in relational life that now moves beyond the family and into elementary school (Siegel, 1999; Ciompi, 1991). Of the four neurobiological regulators outlined by Cozolino (2006), the one I emphasize at this point along our developmental journey is the HPA complex of hormone regulation which organizes puberty. As adolescence approaches, I turn to Harter (2012) and her two perspectives on teen age functioning. Risk taking, so much a part of this stage of life, comes under executive function regulation through the maturation of the prefrontal cortex and its extensive interconnectivity. Second, as emotions become more self-conscious and linked to social morality and guilt processes, biology becomes harnessed with culture (in all its diversity, determinism and imperatives; Lewis, 2003) in the service of sustaining positive relationships and strong friendship, as well as romantic, bonds.

I end this section with a summary of Tronick's Mutual Regulation Model (MRM) as a well-researched example of the psychobiological co-regulation theorizing compatible with my thinking. The MRM rests on two propositions: that human functioning is understood to be organized according to complex dynamic systems and, eschewing abstract theorizing, the MRM is experience near and clinically authentic. As development proceeds through optimal dyadic engagement, a sense of certitude about the self emerges as new information contributes to a joyous sense of expansion and coherence, knowing "my place in the world." Stolorow and Atwood's (1992) concept of the experience of the self-delineating selfobject contributes to the feeling of "my place in the world" as a firm subjective reality.

Resilience is also considered and understood as a feature of regulatory processes. As the infant-mother dyad undergoes innumerable stressful episodes of matching-mismatching-reparation, a gradual sense of mastery emerges in which the infant is able to tolerate better the next episode of stress. In the context of loving relationships the infant develops the capacity for resiliency and greater self-sufficiency.

In Chapter Three, I present a psychotherapy case, that of Myles, as an example both of the unexpected vicissitudes of normal and abnormal development, and, in particular, the significance of the emergence of a self-referencing metaphor, that of "Top dog on the floor," in the course of treatment. In the *Transformational Self*, I generated the hypothesis that adolescence can be said to come to an end with the activation of the transformational neural self, a reconfigured, cohesive state of mind, that presents the late adolescent with a new set of identifications and potentials that can shape subsequent development. At this point let me define the neural self using three sources. First, Damasio (1994, pp. 236–244) defines "The neural self, as a complex adaptive system, is a repeatedly reconstructed biological state that endows our experience with subjectivity and that depends on the continuous reactivation of images about our identity and our body" (Damasio in Greenman, April, 2007, pp. 52–53). Second, from Schore (2002, pp. 443–448) we borrow: "In addition, the neural self emerges as a result of right hemisphere maturation. It is a body self through the gradual differentiation of mutual co-regulation of state and affective experience between the caregiver and the self." And third, Feinberg (2009, p. xi) defines the neural self "essentially by its coherence: *a unity of consciousness in perception and action that persists in time*" (in Bendicsen, 2013, pp. 185–186). Also, the self of self psychology breathes metaphoric life into the neural self. The neural self, then, incorporates dynamic systems thinking, embodied subjective processes, mutual co-regulation among other neural selves and a measure of its tonus or vitality through evaluating its coherence. The Transformational Self (Bendicsen, 2013) is a phase-specific neural self which can become mobilized with the activation of the self-referencing metaphor at the end of adolescence.

The data supporting the Transformational Self hypothesis was generated from two psychotherapy cases of adolescent girls struggling with varying degrees of family dysfunction and psychopathology. Neither girl was psychotic (Bendicsen, 2013). The case of Myles allows me to expand my hypothesis to include in my adolescent cohort, males with psychotic conditions. These three cases, individual statements all, constitute too small a sample to move us toward a universal statement regarding the presence of the Transformational Self at the crossroads of late adolescence and young adulthood. Rather than overgeneralize and lay a claim of certitude, I encourage future human developmental researchers to explore and uncover the presence of yet-to-be-verbalized self-referencing metaphors in this cohort. See comments from Popper at the end of this section.

In Chapter Four I discuss two pathways to adulthood, one fairly straightforward as found among some youth in classical Greece, the other ambiguous, as in some adolescent and emerging adult members of contemporary society. Also, I introduce the metaphor of "Lapham's wound" as a way to characterize the passage from late adolescence to young adulthood, albeit from an academic perspective. At this critical developmental juncture, those adolescents who open themselves up to self-reflection and subsequent personal change have the opportunity to build an enhanced integrated self. The most essential dimension of personal change is the capacity for self-regulation and, with it, the delay of gratification, the hallmark of maturity (Steinberg, 2014, pp. 76–78; 107–124). Experiencing Lapham's wound is understood as a developmental achievement, one that equips late adolescents to transition into functional, reflective young adults. In other words, when combined with the Transformational Self, *the appearance of self-referencing metaphor can evolve into self-regulating metaphor* that channels the identifications and potentials of the late adolescent into stable young adulthood.

I also examine the differences among the dominant configurations of theories: either a single theory, a pluralistic array of theories, eclecticism or an attempt at an integration of theories or an avoidance of integration as in "epigenetic hierarchical arrangement" (Jaffe, 2000). I place a developmental algorithm comfortably in a theoretical pluralistic context.

In Chapter Five, I propose that a displacement of older psychological theories, such as drive theory, needs to occur to be replaced by the following seven elements of a developmental algorithm, a new, contemporary explanatory synergy. In this different paradigm are seven dimensions. The elements will include: 1) embodied metaphor theory, 2) attachment theory, 3) self psychology with intersubjectivity and relational theory, 4) cognition, 5) contemporary psychoanalytic developmental theory, 6) non-linear dynamic systems theory and 7) neurobiology with narrative theory. I maintain that these seven elements, which are applied in my case example, constitute a *compatible set of interlocking theories* and predict that they will become part of a compelling contemporary explanatory system known as regulation theory (Hill, 2010). The goodness of fit amongst these components is optimal in that, not infrequently, they reference each other in the clinical literature. It should be added that the embodiment of human processes from self-referencing metaphor (Lakoff, 2009; Bendicsen, 2013) to intersubjective experience (Merleau-Ponty, 1945) to Panksepp's (1998, 2012) theory of motivation and the origin of the archaic self, is getting better understood. In embodiment processes the body becomes the primary sight of knowing rather than consciousness.

Some of the latest examples of this interlocking theoretical trend may be found in Lucente's *Character Formation and Identity in Adolescence* (2012), Ammaniti and Gallese's *The Birth of Intersubjectivity* (2014) and most recently, Dick and Muller's (editors) *Advancing Developmental Science: Philosophy: Philosophy, Theory and Method* (2017). In the first, the author blends various psychodynamic and psychoanalytic theories with neuroscience findings to explicate the therapeutic process. In the second, the authors attempt to weave a coherent explanatory narrative amongst the domains of intersubjectivity, neurobiology, infant development and psychoanalysis to explain how the simultaneous brain development between mother and infant induces alterations in the emerging subjectivities of both. In the third example, the editors weave concepts from embodiment, complexity theory and neuroscience, among other domains of knowledge, into a relational perspective on development.

In addition, I present a discourse, a compare/contrast elaboration on the relationship among self psychology (in particular the seminal concept of the self-selfobject), intersubjectivity theory and relational psychoanalysis. This theoretical journey serves as a justification for including subjective experience in my developmental algorithm and anticipating criticism for mixing objective concepts with subjective experience.

Next, I open by providing a parallel case formulation using a regulation theory perspective to demonstrate the enhanced explanatory power a set of compatible interlocking theories has over a single theory. I operationalize regulation theory by using my seven-dimensional developmental algorithm. Each dimension contributes toward the explanatory synergy today's clinical environment requires. I next present two new parallel case formulations, the social brain perspective and the non-linear dynamic systems perspective. These are designed to show the reader the intriguing possibilities for tomorrow's clinical explanatory frameworks. Each finds an easy fit into the continuously reconfigured developmental algorithm.

I also reflect on the multiple conceptualizations of supportive relationships including: Goldstein's "ego allies," Kohut's "selfobject," Newman's "usable object" and Galatzer-Levy and Cohler's "essential other."

Today's clinical developmental professionals resonate with my move away from the traditional overreliance of task attainment or completion to assess closure to the adolescent phase. Instead, I present a set of developmental processes to challenge our understanding of the forces at play that move us through this point in the life cycle. These seven developmental processes are:

A. Those associated more with evolutionary/biological maturational forces that signal advanced levels of differentiation and reorganization:

1) the need for attachment (interconnectedness) – differentiation/individuation (Lyons-Ruth, 1991; Doctors, 2000; Palombo, 2016);

2) the need for cognitive dissonance to resolve opposing, conflicting biological states into one coherent schema (Bendicsen, 2013);
3) the mobilization of the Transformational Self exerting the force of a dynamical attractor (Bendicsen, 2013);
4) the attenuation of the sense of grandiosity that accompanies the Transformational Self (Bendicsen, 2018); and
5) the need to consolidate the reconfigured self-state with a sense of certitude (Tronick, 2007) in a new reality about being a distinctive subjectivity in an intersubjective environment (Stolorow and Atwood, 1992).

B. Those associated more with philosophical/cultural developmental forces that serve as imperatives in the creation of the capacity for resilience (defined as successful adaptation in the face of adversity) necessary for healthy functionality in contemporary society:

6) the desirability of living a life organized into the future (Summers, 2013);
7) the differentiation of mutual recognition between two subjectivities, self and other (Benjamin, 1990), which is a process for the lifespan that needs to be renegotiated at every developmental juncture.

These seven processes exert a dynamism (from Sullivan, 1953, p. 103, "the relatively enduring pattern of energy transformations which recurrently characterize the organism in its duration as a living organism," in Palombo, Bendicsen and Koch, 2009, pp. 209, 231) of action potential, an ontological force that collectively contributes to pulling the late adolescent into young adulthood. These processes unfold neither sequentially nor comprehensively, but rather some or all may manifest according to organizational dynamics commensurate with complexity theory. As the new (one or more) self-referencing metaphor(s) evolves into the self-regulating metaphor(s) these processes become channelized. In this experience the adolescent feels a coherent sense of direction emerge which guides and counterbalances the vulnerability the late adolescent feels in these uncharted waters and emboldens the adolescent into risk-taking identifications.

In Chapter Six, I present a brief case formulation of Myles from a traditional psychodynamic-psychoanalytic self-psychological perspective. It is organized according to the Perry, Cooper and Michels' (1987) four-part Psychodynamic Formulation Model. Its four parts are: Summarizing Statement, Description of Nondynamic Factors, Psychodynamic Explanation and Predicting Response to the Therapeutic Situation. As I consider other perspectives, the Cooper formulation will assure uniformity and provide for an optimal opportunity for a compare/contrast exercise.

In Chapter Seven, I outline the best practice standards for the treatment of schizophrenic spectrum disorders. Most therapists develop specializations with a certain circumscribed patient population in order to exploit heightened levels of competence and skill in diagnosis and treatment. In cases where referral to such a specialist is obviated by a long-standing therapeutic alliance with the patient, the therapist must get up to speed with best practice standards and consultation. In the case presented, a three-pronged effort has been employed using psychotropic medication, a nurturing, loving, psychoeducational selfobject milieu and supportive, reflective psychotherapy with consultation.

In Chapter Eight, I reflect on the adolescent passage from the perspective of Arnett's (2004) emerging adulthood proposal and the sociological phenomenon of the Millennials. I maintain that Erikson's (1950) framework contains sufficient explanatory power to obviate the need for a new developmental phase, "emerging adulthood." However, as a description of the millennials, the phrase, emerging adulthood, is very useful. As one studies the millennial cohort, one dimension clearly distinguishes it from the previous two generations. With the economic downturn, increasing college loan debt and with the greater numbers of emerging adults living with parents and relatives, it is certain that adolescence, as a developmental phase, will continue to expand.

In Chapter Nine, I summarize my outline and, in so doing, hope to persuade my readers of the logic of my hypothesis: that regulation theory, expressed and operationalized in my developmental algorithm, is the preferred contemporary *zeitgeist*,[1] allowing clinicians to combine neuroscience research findings into an enlarged explanatory synergy with other domains of knowledge. The explanatory framework is an example of pluralistic theory construction. It assumes a unitary dynamic in that there is an internal coherence bringing harmony to the diverse theoretical sub-orientations. A developmental line for the late adolescent is suggested in which the emergence of one or more self-referencing metaphor(s) evolves into a self-regulating metaphor leading the way to personal metamorphoses.

In today's rapidly evolving clinical world, the effective clinician must practice in a biopsychosocial context. Empirical data demonstrates that

> much of mental life is unconscious, that social forces in the environment shape the expression of genes, and that the mind reflects the activity of the brain. We now practice in a situation of "both/and" rather than "either/or." Although it is true that all mental functions ultimately are products of the brain, it does not follow that the biological explanation is the best or most rational model for understanding human behavior. Contemporary neuroscience does not attempt to reduce everything to genes or biological entities. Well-informed neuroscientists focus on an integrative rather than a reductive approach and recognize that psychological data are just as valid scientifically as biological findings.
> (Cloninger, 2004 and LeDoux, 2012 in Gabbard, 2014, p. 4)

Because the domain of mind and the domain of brain have different languages, the modern clinician must strive to be bilingual, mastering both languages in order to know the whole person and provide optimal patient care (Gabbard, 2014, p. 7).

As we create new propositions and hypotheses we should be mindful of Freud's encouragement in the formulation of theory construction to adopt a playful attitude and let the attendant associational process unfold (Freud, 1923, p. 15). Theoretical playfulness or imaginative creativity is a function of the yield from interesting client data. Further advances in the development of my hypothesis will have to await the arrival of additional case material. Erikson credited his theoretical creativity directly to the intellectual stimulation he derived from client treatment opportunities (Friedman, 1999, pp. 260–286).

I have one last thought. The case of Myles is the third case example to illustrate the significance of the self-referencing metaphor in late adolescence. As the Transformational Self forms, the self-referencing metaphor ("Top dog on the floor") or quasi-metaphor ("I keep the rhythm") have the potential to evolve into self-regulating metaphors that serve to channel and maintain new identifications and capacities for the young adult. The theoretical basis to the Transformational Self is derived from data obtained from three psychotherapies and so is an example of inductive logic.[2]

A cautionary comment is in order at this point. Popper (1935, pp. 93–111), who was dedicated to the study of epistemology and the growth of knowledge, believed that empirical reasoning is reducible to sensory perception. The ensuing statements of experience are perceptual knowledge and form the basis for truth about the world: "it is the systematic presentation of our immediate convictions" (p. 94). Popper was concerned that the compelling process of induction leads inexorably from individual statements to universal statements, creating a false sense of certitude about our conclusions. He embraced the term psychologism (Erdmann, 1866) to mean "the doctrine that statements can be justified not only by statements, but by perceptual experience" (p. 94). To counter the tendency to overgeneralize or universalize, he privileged deductive reasoning, but because I generated the Transformational Self through an inductive reasoning process, I am mindful that I cannot claim Transformational Self universality for all late adolescents. Until more data is gathered, I say for now that some, not all, late adolescents will experience the instantiation of the Transformational Self and experience a more definitive transition into young adulthood.

At this point, I shall have to content myself with the progress to date. In closing, let me cite two luminaries in the field of theory construction, Albert Einstein and Karl Popper.

> Albert Einstein, in his autobiography, defined his criteria for a serviceable theory: "A theory is more impressive, the greater the simplicity of its premises, the more kinds of things it relates, and the more extended the area of its applicability." A workable theory, therefore, encompasses and accounts for the widest range of human behavior, normal and abnormal.
> (A. Einstein/P. Schlipp, 1991, in Greenman, April, 2007)

Karl Popper wrote:

> The empirical basis of objective science has thus nothing "absolute" about it. Science does not rest upon solid bedrock. The bold structure of its theories rises, as it were, above a swamp. It is like a building erected on piles ... and if we stop driving the piles deeper, it is not because we have reached firm ground. We simply stop when we are satisfied that the piles are firm enough to carry the structure, at least for the time being.
>
> (Popper, 1935, p. 111)[3]

Notes

1 The general intellectual, moral and cultural climate of an era, *Webster's New Collegiate Dictionary*, 1981, p. 1352.
2 With a grasp of the seven developmental processes I have formulated, it is possible to understand better Mark Twain's thinking about his passage through adolescence: "When I was a boy of fourteen, my father was so ignorant I could hardly stand to have the old man around. But when I got to be twenty one I was astonished at how much he had learned in seven years." (This saying first appeared in the Readers Digest, September 1937 and its attribution to Samuel Langhorne Clemens is questionable.)
3 "The old scientific ideal of *episteme* – of absolutely certain, demonstrable knowledge – has proven to be an idol. The demand for scientific objectivity makes it inevitable that every scientific statement must remain *tentative forever*. It may indeed be corroborated, but every corroboration is relative to other statements which, again, are tentative. Only in our subjective experiences of conviction, in our subjective faith, can we be 'absolutely certain.'" (Popper, 1935, p. 280).

> Science never pursues the illusory aim of making its answers final, or even probable. Its advance is, rather, towards the infinite yet attainable aim of ever discovering new, deeper, and more general problems, and of subjecting its ever tentative answers to ever renewed and ever more rigorous testing.
>
> (Popper, 1935, p. 281)

Appendix I
CRITICAL THINKING MENTAL HEALTH DECISION-MAKING FLOW CHART

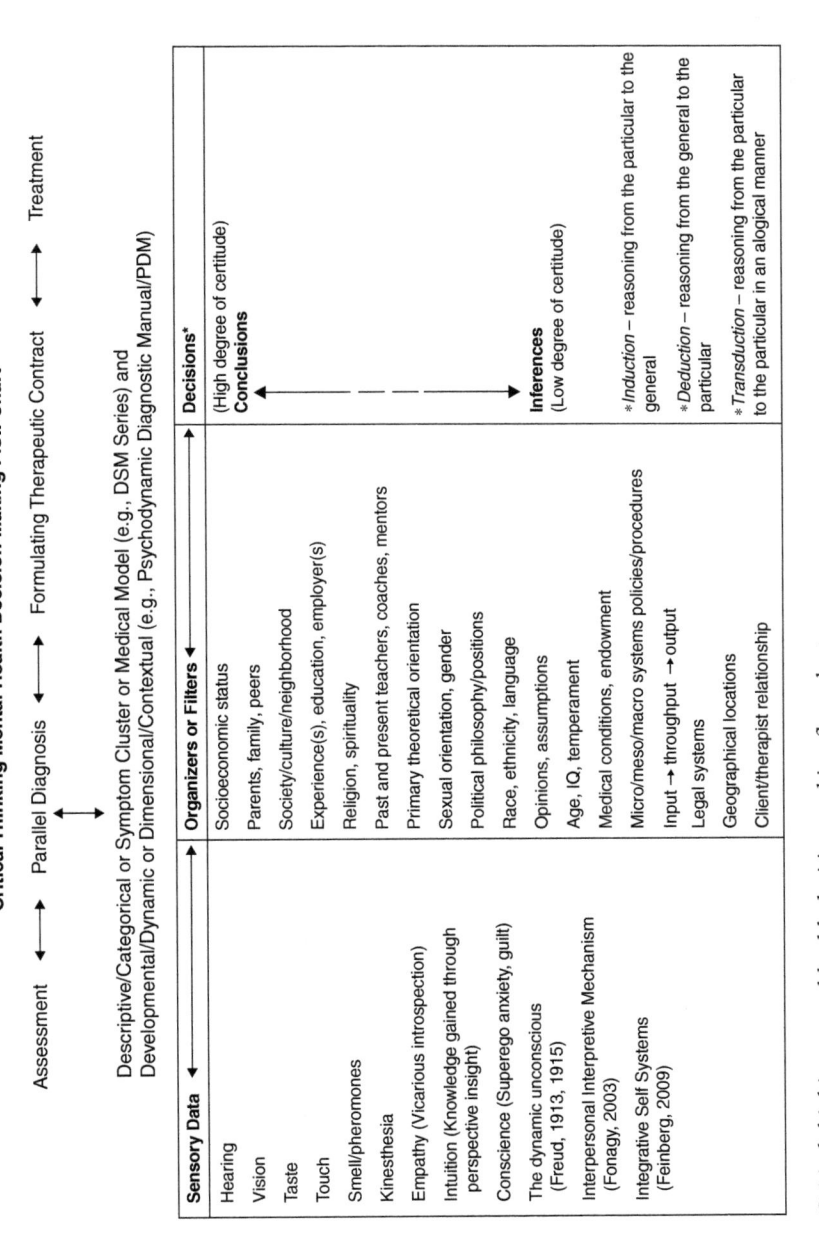

Critical Thinking Mental Health Decision-Making Flow Chart

Assessment ⟷ Parallel Diagnosis ⟷ Formulating Therapeutic Contract ⟷ Treatment

Descriptive/Categorical or Symptom Cluster or Medical Model (e.g., DSM Series) and
Developmental/Dynamic or Dimensional/Contextual (e.g., Psychodynamic Diagnostic Manual/PDM)

Sensory Data	Organizers or Filters	Decisions*
Hearing	Socioeconomic status	(High degree of certitude)
Vision	Parents, family, peers	**Conclusions**
Taste	Society/culture/neighborhood	
Touch	Experience(s), education, employer(s)	
Smell/pheromones	Religion, spirituality	
Kinesthesia	Past and present teachers, coaches, mentors	
Empathy (Vicarious introspection)	Primary theoretical orientation	
Intuition (Knowledge gained through perspective insight)	Sexual orientation, gender	⟷
Conscience (Superego anxiety, guilt)	Political philosophy/positions	
The dynamic unconscious (Freud, 1913, 1915)	Race, ethnicity, language	**Inferences**
Interpersonal Interpretive Mechanism (Fonagy, 2003)	Opinions, assumptions	(Low degree of certitude)
Integrative Self Systems (Feinberg, 2009)	Age, IQ, temperament	
	Medical conditions, endowment	*Induction* – reasoning from the particular to the general
	Micro/meso/macro systems policies/procedures	
	Input → throughput → output	*Deduction* – reasoning from the general to the particular
	Legal systems	
	Geographical locations	*Transduction* – reasoning from the particular to the particular in an alogical manner
	Client/therapist relationship	

Critical thinking mental health decision-making flow chart

Appendix II
THEORIES OF ALCOHOLISM

Disease-biological model	Moral model	Psychodynamic-psychoanalytic model	Psychosocial theory of alcoholism
The DB model was introduced in 1952 by E. M. Jellinek who studied the signs and symptoms of 2,000 AA members. He standardized the disease process by suggesting the course consists of four phases: 1. Prealcoholic – individual consumes alcohol in social situations, getting psychological relief from the alcohol, develops tolerance and gravitates towards situations where alcohol is present; 2. Prodromal – individual becomes sneak drinker, experiences blackouts, guilt over self-destructive behavior becomes intense, heavy use of denial and minimization and struggles hard to appear normal; 3. Crucial – individual now experiences loss of control, makes excuses/uses rationalization for inappropriate behavior, attempts abstinence, varies alcoholic drinks, has driving-related accidents and uses eye-opener to manage physical dependence; 4. Chronic – individual encounters grave employment, family and social consequences, intoxication occurs daily, severe tolerance occurs, potentially fatal medical conditions (such as cirrhosis of the liver, Wernicke's encephalopathy, Korsakoff's psychosis and/or alcohol dementia), alibi system fails with unexplained anxiety.	The M model is rooted in "John Barleycorn," an English folksong with ancient derivation. JB in the song is a personification of the important cereal crop barley, and the alcoholic beverages made from it, beer and whisky. JB suffers attacks, indignities and death, representing and corresponding to the various stages of barley cultivation such as reaping and malting. Jack London published an autobiographical novel under the same name in 1913. M model traces its official beginning to the 1820s with the founding of the temperance movement in England/Ireland. In the early 1800s in the US a pledge of abstinence was taken by preachers led by J. B. Gough. In the late 1800s the temperance movement merged efforts with social justice/civil rights issues, viz. poverty, suffrage/women's rights, education reform and slavery. Pioneers were Susan B. Anthony (1820–1906). who helped lead Women's Christian Temperance Union in 1870s and the Anti-Saloon League in 1895; Carrie Nation (1846–1911), who attacked saloons singing hymns, was arrested 30 times; Frances Willard (1839–1898), who helped found WCTU in 1874,	The PP model is regarded with skepticism by mental health professionals and society because of the small impact interpretations of unconscious motivations have had on drinking behavior. Psychological needs are addressed and facilitated by AA organization participation with lasting structural personality changes. The PP model helps facilitate an understanding of some of the changes rendered by the AA approach. Abstinence in a community with other sufferers assist the individual with management, impulse control and other ego functions. AA works well in an individual who can accept the idea that he/she has no control over their drinking and needs to surrender to a higher power and for those free of other psychiatric disorders. No specific known personality traits are predictive, but psychoanalytic observers have noted structural defects such as ego weakness and difficulty maintaining self-esteem. According to self psychology and object relations theories alcohol is said to replace missing psychological structures and restores self-regard and inner harmony, although it only lasts as long as the intoxication. For example, in self psychology the selfobject is experienced as performing one or more functions such as calming, regulating, and/or soothing.	In the PS model alcohol use and alcoholism are best viewed as end products of a complex combination of biopsychosocial influences. Hypothesis: "in childhood, biologically based vulnerabilities in emotional and behavioral regulation (temperament or personality) interact with poor parenting to create emotional distress and exposure to negative peer influences, both of which create alcohol misuse." In families with a history of alcoholism there is a higher prevalence of psychopathology (mental and behavioral disorders, viz., antisocial personality and depression), more adverse family environments and physiologic responses to alcoholism that are known to be associated with risk – in particular, a lack of sensitivity to alcohol's intoxicating effects or an increased sensitivity to its anxiety-reducing effects. Mediation models measure the effects of risk factors over time. In the Deviance Proneness model, the risk for alcohol misuse is part of a larger context of poor socialization and adolescent problem behavior. In the Negative Affectivity model, intergeneration transmission of alcoholism exposes children to high levels of stress and children are temperamentally hyper-reactive to stress. In the Sensitivity to the Effects of Alcohol model, children of alcoholics have greater sensitivity to the stress response-dampening effects of alcohol and less sensitivity to the negative effects of alcohol on the body such as swaying and intoxication.

(*continued*)

Disease-biological model	Moral model	Psychodynamic-psychoanalytic model	Psychosocial theory of alcoholism
Following Jellinek's work, the American Psychiatric Association began to use the term "disease" to describe alcoholism in 1965. The American Medical Association followed in 1966. The DB model is now the dominant and commonly accepted explanation for alcoholism.

Alcoholism is not the person's fault, it is a biological disease, like diabetes. Promoted by the 12 Step AA organization, but in contrast to the DB model, AA expects the individual to take responsibility for their alcoholism.

Individuals with alcoholism have an inherent predisposition to addiction to exogenous substances.

Psychological factors are irrelevant.

The DB model gained support from genetic studies of substance-related disorders, e.g. children raised apart from alcoholic parents still have an increased chance of susceptibility.

Twin studies of both male and female pairs suggest that genetic factors also play a major part in alcoholism, alcohol abuse and alcohol dependence.

The origins of alcohol disease are believed to be related to specific genetic risks for substance-use disorders in addition to environmental influences. | and was attacked by Ida B. Wells for racist attitudes. Their anti-alcoholism views reinforced the M model and led to the passing of the 18th Amendment known as Prohibition, in 1919 and repealed in 1933.

In the late 1800s the temperance movement merged efforts with social justice/civil rights issues, viz. poverty, suffrage/women's rights, education reform and slavery. Pioneers were Susan B. Anthony (1820–1906), who helped lead Women's Christian Temperance Union in 1870s and the Anti-Saloon League in 1895; Carrie Nation (1846–1911), who attacked saloons singing hymns, was arrested 30 times; Frances Willard (1839–1898), who helped found WCTU in 1874, and was attacked by Ida B. Wells for racist attitudes. Their anti-alcoholism views reinforced the M model and led to the passing of the 18th Amendment known as Prohibition, in 1919 and repealed in 1933.

In 1865 in London's East End, Methodist minister Wm. Booth and wife Catherine started the Salvation Army. In 1880 the SA began work in the US assisting alcoholics, morphine addicts, prostitutes and other "undesirables" unwelcomed into polite Christian society. | Others also observed that alcoholic patients have problems with self-esteem, the modulation of affect and self-care.

In studies of alcoholics, in whom an attempt was made to diagnose personality disorder, it was found that the prevalence of comorbid Axis II conditions varied from 14%–78%.

Caution: Persons labeled with DSM disorders need to be considered in the context of the total person.

After sobering up, they suffer depression. Depression caused by use may arise from the painful recognition that they have hurt others. They may also mourn valuable things that have been lost or destroyed as a result of their addictive behavior.

When depression and alcoholism are found together, the synergistic or additive effect can result in a high level of suicidality.

Most therapists argue that abstinence is necessary for psychotherapy to be effective.

In PP drive and ego theory, sources of alcoholism may relate to seeking sensuous satisfaction, conflict amongst personality components and/or psychosexual fixations. | In the contribution from neuroscience, poor executive function deficits may predict increases in alcohol consumption among young adults with a history of alcoholism.

In terms of life span psychosocial development, antecedents to alcoholism can be seen in the preschool years in the form of deficits in self-regulation, emotional reactivity and conduct problems.

In high-risk families, prevalence of alcohol use increases greatly after eighth grade, along with delinquency and risky sexual behavior.

Central motivations for alcoholism are stress reduction, relief from anxiety and mood enhancement.

In the role of cognition in alcoholism, Explicit Beliefs and Expectations pertain to conscious expectancies which influence the experience of alcohol use, perhaps in the manner of self-fulfilling prophecies. Implicit Beliefs and Expectations pertain to unconscious expectancies which activate motivational states outside of the individual's awareness. Both Explicit and Implicit Beliefs and Expectations may serve as predictors of alcoholism through deliberate or spontaneous behaviors, respectively. |

Vaillant (1983) found that the only reliable predictor of alcoholism was antisocial behavior. Depression, anxiety and other psychological characteristics are consequences, rather than causes, of the disorder. The DB model, along with self-help groups, has been less effective when applied to drug abusers because of differences between alcoholics and poly-drug users that require differential approaches. (See PP model.) Narrow interpretation of this model may cause clinicians to ignore how these factors contribute to relapse in the course of the illness. Vaillant's controversial approach differs from the DB's de-psychologizing of alcoholism. He believes personality factors, like low self-esteem (e.g. the inability to self-care), identify a critical feature in the alcoholic person's vulnerability. Because it is believed there is a spectrum of individuals struggling with alcoholism, it is difficult to generalize about personality vulnerabilities (Vaillant, 1983); (Gabbard, 1994, pp. 361–365).	The M model views alcoholic individuals as bearing complete responsibility for their alcoholism. Alcoholics are selfish, hedonistic people interested only in pursuit of their own pleasure with no regard for the feelings of others. The M model has a fundamentalist religious belief that alcohol is a sign of moral turpitude (conduct that is contrary to community standards of good morals; it is behavior that is considered shameful, base, depraved). The individual has failings of will power which are linked to notions of sin and punishment. Such behavior is best handled through the legal system with the application of appropriate consequences for alcoholic individuals. Now more frequently applied to drug abusers than alcoholics mainly because of the overlap between crime and drug abuse (Gabbard, 1994, pp. 360–361).	In earliest PP theory, alcoholism is associated with the psychosexual developmental organization and infantile arrests or fixations. In the oral fixation, alcoholism is manifested by an id obsession with stimulating the mouth. Drinking is a sedative and source of warmth. Adult personality traits include dependency, inability to postpone gratification and impulsivity. Sublimations include smoking, gum chewing, excessive talking, etc. In anal fixation, alcoholism is manifested by conflict between the id (rebellion against control) and the superego (guilt over inability to maintain self-restraint), resulting in ambivalent behavior with repeated, compulsive drinking alternating with guilt and compulsive efforts to abstain. In phallic fixation, alcoholism is manifested in a compromised Oedipal (or Electra) complex where, in the adult, due to superego lacunae, sexual partners are attacked or sexual partners are sought who are taboo. In the genital stage, the ego has reached optimal adaptive control and harmony amongst the id, ego and superego is maintained. Also, multiple fixations are possible with oral passivity alternating with anal compulsiveness and phallic competitiveness and aggressivity (Gabbard, 1994, pp. 361–365); (Barry, 1988, pp. 103–141).
		Research on psychosocial factors in alcohol use/abuse encompasses the investigation of the effects of multiple, interactive risk and protective factors (*Ninth Report to the US Congress on Alcohol and Health*, 1997, pp. 181–191). *Theories of Alcoholism*

REFERENCES

Abraham, K. (1927). "A short study of the development of the libido, viewed in the light of mental illness." In Ernest Jones (ed.), *Selected Papers of Karl Abraham* (1949). London, UK: The Hogarth Press.

Agosta, L. (2014). *A Rumor of Empathy: Rewriting Empathy in the Context of Philosophy*. New York: Palgrave Macmillan.

Aleman, A. (2014). "Neurocognitive basis of schizophrenia: Information processing abnormalities and clues for treatment." *Advances in Neuroscience*. Article ID 104920. Netherlands: University of Groningen.

American Psychiatric Association. (April, 2004). "Practice guidelines for the treatment of patients with schizophrenia." Second edition. *American Journal of Psychiatry*. 161: 1–56.

American Psychiatric Association. (September, 2009). "Guideline watch: Practice guideline for the treatment of patients with schizophrenia." http://psychiatryonline.org/content.aspx?bookid+28&sectionid+1682213.

Amit, D. J. (1989). *Modeling Brain Function: The World of Attractor Networks*. Cambridge: Cambridge University Press.

Ammaniti, M. and Gallese, V. (2014). *The Birth of Intersubjectivity: Psychodynamics, Neurobiology, and the Self*. New York: W. W. Norton & Company.

Andreasen, N. C. (2001). *Brave New Brain: Conquering Mental Illness in the Era of the Genome*. New York: Oxford University Press.

Applegate, J. (Fall, 1989). "Mahler and Stern: Irreconcilable differences?" *Child and Adolescent Social Work*. 6(3).

Arieti, S. (1974). *Interpretation of Schizophrenia*. Second edition. New York: Basic Books.

Arnett, J. (2004). *Emerging Adulthood: The Winding Road from the Late Teens Through the Twenties*. Oxford: Oxford University Press.

Aron, I. (1991). "The patient's experience of the analyst's subjectivity." *Psychoanalytic Dialogues*. 1:29–51.

Aron, I. (1996). *A Meeting of the Minds: Mutuality in Psychoanalysis*. Hillsdale, NJ: The Analytic Press.

Aron, I. (2005). *Relational Psychoanalysis (Vol. 2) Innovation and Expansion*. Hillsdale, NJ: The Analytic Press.

Aron, I. and Harris, A. (1993). *The Legacy of Sandor Ferenczi*. New York: Analytic Press.

Arnsten, A. F. T. (1998). "The biology of being frazzled." *Science*. 280: 1711–1712.
Aserinsky, E. and Kleitman, N. (1953). "Regularly occurring periods of eye motility and concurrent phenomenon during sleep." *Science*. 118: 273.
Aserinsky, E. and Kleitman, N. (1955). "Two types of ocular motility during sleep." *Journal of Applied Physiology*. 8:1.
Atwood, G. and Stolorow, R. (1984). *Structures of Subjectivity*. Hillsdale, NJ: The Analytic Press.
Bacal, H. (1985). "Optimal responsiveness and the therapeutic process." In A. Goldberg (ed.), *Progress in Self Psychology*. Vol 1. New York: Guilford Press, pp. 202–226.
Barratt, B. B. (2013). *What is Psychoanalysis: 100 Years After Freud's Secret Committee*. London, UK and New York: Routledge.
Barry, H. (1988). *Psychoanalytic Theory of Alcoholism*. C. D. Chaudron and D. A. Watson (eds). Toronto: Alcoholism and Drug Addiction Research Foundation, pp. 103–144.
Barton, R. I. M. (1998). "The social brain hypothesis." *Evolutionary Anthropology*. 6(5): 178–190.
Barton, R. A. and Dunbar, R. L. M. (1997). "Evolution of the social brain." In A. Whiten and R. Byrne (eds), *Machiavellian Intelligence*. Vol. II. Cambridge, UK: Cambridge University Press.
Basch, M. F. (1980). *Doing Psychotherapy*. New York: Basic Books.
Baxter, R. D. and Liddle, P. F. (1998). "Neuropsychological deficits associated with schizophrenic syndromes." *Schizophrenic Response*. 30: 239–249.
von Bertalanffy, L. (1952). *Problems of Life*. New York: Wiley.
von Bertalanffy, L. (1968). *General Systems Theory*. New York: Braziller.
Belsha, K. and Girardi, L. (March 12, 2015). "Aurora school stunned by fatal accident: 2 teens killed in crash with semi: Most of the people … don't know what to say." Chicago, IL: *Chicago Tribune*.
Bendicsen, H. K. (August 1992). "Achieving the capacity to tolerate ambiguity: The role played by literature in the psychotherapies of three late adolescents." *The Association of Child Psychotherapists Bulletin*. Vol. 9. Chicago, IL: The Association of Child Psychotherapists Annual.
Bendicsen, H. K. (2013). *The Transformational Self: Attachment and the End of the Adolescent Phase*. London, UK: Karnac Press.
Benjamin, J. (1988). *The Bonds of Love: Psychoanalysis, Feminism, and the Problem of Domination*. New York: Pantheon.
Benjamin, J. (1990/1995). "An outline of intersubjectivity: The development of recognition." *Psychoanalytic Psychology*. 7: 33–46.
Benjamin, J. (1995/1998). *Like Subjects, Love Objects: Essays on Recognition and Sexual Difference*. New Haven, CT: Yale University Press.
Berman, M. (March 15, 2017). "Are leaks from fount of youths?: Ex-CIA chief thinks millennials' values could mean they are." Chicago, IL: *Chicago Tribune*.
Bernstein, R. J. (1983). *Beyond Objectivism and Relativism: Science Hermeneutics and Praxis*. Philadelphia, PA: University of Pennsylvania Press.
Berzoff, J., Flanagan, L. M. and Hertz, P. (2008). *Inside Out and Outside In: Psychodynamic Clinical Theory and psychopathology in Contemporary Multicultural Contexts*. Second edition. Lanham, MD: Jason Aronson.
Bildungsroman. Encyclopedia Britannica Online, December 10, 2015. www.britannica.com/art/bildungsroman.
Blos, P. (1962). *On Adolescence: A Psychoanalytic Interpretation*. New York: Free Press.
Blos, P. (1967). "The second individuation process of adolescence." *Psychoanalytic Study of the Child*. 22: 162–186.

References

Boak, J. (December 6, 2015). "Recovery divided on age, race, place: Millennials find themselves priced out of housing market." Chicago, IL: *Chicago Tribune*.

Bollas, C. (1987). *The Shadow of the Object: Psychoanalysis of the Unknown Thought*. London, UK: Free Association Press.

Bollas, C. (2015). *When the Sun Bursts: The Enigma of Schizophrenia*. New Haven, CT: Yale University Press.

Bomkamp, S. (November 2, 2015). "Millennials skipping aisles of 'big food' at grocery stores." Chicago, IL: *Chicago Tribune*.

Bowen, A. (November 1, 2015). "Birthed in the digital age." Chicago, IL: *Chicago Tribune*.

Bosanquet, B. (1900). *The Education of the Young in Plato's Republic*. London, UK: J. C. Clay and Sons, pp. 8–9.

Bowlby, J. (1969). *Attachment and Loss. Volume I. Attachment*. New York: Basic Books.

Bowlby, J. (1973). *Attachment and Loss. Volume II. Separation, Anxiety and Danger. Attachment*. New York: Basic Books.

Bowlby, J. (1980). *Attachment and Loss. Volume III. Loss: Sadness and Depression*. New York: Basic Books.

Brazelton, T. B. (1982). "Joint regulation of neonate-parent behavior." In E. Tronick (ed.), *Social Interchanges in Infancy: Affect, Cognition and Communication*. Baltimore, MD: University Park Press.

Brenner, C. (1973). *An Elementary Textbook of Psychoanalysis*. New York: Anchor Books/Double Day.

Bromberg, P. M. (1998). *Standing in the Spaces: Essays on Clinical Process, Trauma and Dissociation*. Hillsdale, NJ: The Analytic Press.

Bromberg, P. M. (2003). "Something wicked this way comes: trauma, dissociation, and conflict: The space where psychoanalysis, cognitive science, and neuroscience overlap." *Psychoanalytic Psychology*. 20: 558–574.

Brooks, D. (September 8, 2014). "Becoming a Real Person." *The New York Times*.

Brothers, L. (1990). "The social brain: A project for integrating primate behavior and neurophysiology in a new domain." *Concepts Neuroscience*. 1: 27–251.

Cambiano, G. (1995). "Becoming an adult." In Jean-Pierre Vernat, *The Greeks*. Chicago, IL and London, UK: University of Chicago Press.

Cancino, A. (January 27, 2015). "Passionate millennials are new face of activism." Chicago, IL: *Chicago Tribune*.

Cancino, A. (May 17, 2015). "Fast casual still surging." Chicago, IL: *Chicago Tribune*.

Cano, R. C. (February 28, 2017). "Appealing to millennials, Vegas gets e-sports arena." Chicago, IL: *Chicago Tribune*.

Cassidy, D. C. (1992). *Uncertainty: The Life and Science of Werner Heisenberg*. New York: W. H. Freeman and Company.

Catholic Encyclopedia (1913)/Inquisition. Wikisource, October 5, 2013 http://en.wikisource.org/wiki/ Catholic_Encyclopedia (1913)/Inquisition.

Cha, A. E. (July 2, 2017). "As national birthrate falls, will crisis be born?" Chicago, IL: *Chicago Tribune*.

Channick, R. (July 1, 2015). "Millennials buying into Chicago market." Chicago, IL: *Chicago Tribune*.

Chapman, S. (May 20, 2018). "Why millennials are drawn to socialism." Chicago, IL: *Chicago Tribune*.

Chew, G. F. (1968). "Bootstrap: A scientific idea." *Science*. 2: 762–765.

Chew, G. and Ciompi, L. (1991). "Affects as central organizing and integrating factors: A new psychosocial/biological model of the psyche." *British Journal of Psychiatry*. 159: 97–105.

Claridge, G., Pryor, R. and Watkins, G. (1998). *Sounds from the Bell Jar: Ten Psychotic Authors.* Cambridge, MA: Malor Books.

Clark, P. and Greenfield, R. (October 13, 2017). "Millennials with families moving back to the suburbs." Chicago, IL: *Chicago Tribune*.

Clarke, B. L. (1992). "Rene Descartes." In Ian P. McGreal (ed.), *Great Thinkers of the Western World*. New York: HarperCollins.

Cloninger, C. R. (2004). *The Silence of Well-Being: Biopsychosocial Foundations*. Oxford, UK: Oxford University Press.

Colarusso, C. A. (1992). *Child and Adult Developmental: a Psychoanalytic Introduction for Clinicians*. New York: Plenum Press.

Cooper, A. (2008). "American psychoanalysis today: a plurality of orthodoxies." *The Journal of the American Academy of Psychoanalysis and Dynamic Psychiatry*. 36: 235–253.

Coplan, J. D. and Lydiard, R. B. (1998). "Brain circuits in panic disorder." *Biological Psychiatry*. 44: 1264–1276.

Costandi, M. (2006). "The Discovery of the Neuron." http://neurophilosophy.wordpress.com/2006/08/29/the-discovery-of-the-neuron.

Cottrell, L. S. (May, 1978). "Harry Stack Sullivan and George Herbert Mead: An unfinished synthesis." *Psychiatry*. 41: 151–163.

Cozolino, L. (2006). *The Neuroscience of Human Relationships: Attachment and the Developing Social Brain*. New York and London, UK: W. W. Norton & Company.

Cozolino, L. (2010). *The Neuroscience of Psychotherapy: Healing the Social Brain*. Second edition. New York and London, UK: W. W. Norton & Company.

Craveri, M. (1967). *The Life of Jesus*. New York: Grove Press, Inc.

Cross, M. (February 1, 2016). "Should young adults accept financial help from their parents?" Chicago, IL: *Chicago Tribune*.

Crossan, J. D. (1994). *Jesus: A Revolutionary Biography*. San Francisco, CA: Harper.

Curtis, R. C. (2008). *Desire, Self, Mind and the Psychotherapies: Unifying Psychological Science and Psychoanalysis*. Lanham, MD: Aronson.

Dahl, R. E. (2004). "Adolescent brain development: A period of vulnerabilities and opportunities." In R. E. Dahl and L. P. Spear (eds), *Adolescent Brain Development: Vulnerabilities and Opportunities*. New York: New York Academy of Sciences, pp. 1–22.

Damasio, A. R. (1994). *Descartes Error: Emotion, Reason and the Human Brain*. New York: Grosset/Putnam.

Davies, J. M. (1994). "Love in the afternoon: A relational consideration of desire and dread in the countertransference." *Psychoanalytic Dialogues* 4: 153–170.

Davis, S. M. (2002). "The relevance of Gerald Edelman's theory of neuronal group Selection and nonlinear dynamic systems for psychoanalysis." *Psychoanalytic Inquiry*. 22: 814–840.

Deco, G. and Rolls, E. T. (2005). "Attention, short-term memory, and action seeking: A unifying theory." *Progress in Neurobiology*. 76: 236–256.

Deadwyler, S. A. (1987). "Evoked potentials in the hippocampus and learning." In G. Adelman (ed.), *Encyclopedia of Neuroscience*. Vol. I. Boston, MA: Birkhauser, pp. 411–412.

Delbanco, A. and Delbanco, T. (March 20, 1995). "AA at the crossroads." New York: *The New Yorker*.

Dement, W. and Kleitman, N. (1957a). "Cyclical variations in EEG during sleep and their relation to eye movements, body motility and dreaming." *Electroencephalography and Clinical Neurophysiology*. 9: 673.

Dement, W. and Kleitman, N. (1957b). "The relation of eye movements during sleep to dream activity: An objective method for the study of dreaming." *Journal of Experimental Psychology*. 53: 89.

Deresiewicz, W. (August 4, 2014). "I saw the best minds of my generation destroyed by the Ivy League: Against the tyranny of elite education." New York: *The New Republic*.
Deresiewicz, W. (2014). *Excellent Sheep: The Miseducation of the American Elite and the Way to a Meaningful Life*. New York: Free Press, pp. 84–85.
Diagnostic and Statistical Manual of Mental Disorders III. (1980). Third Edition. Washington, DC: American Psychiatric Association.
Diagnostic and Statistical Manual of Mental Disorders 5. (2013). Fifth Edition. Arlington, VA: American Psychiatric Association.
Dick, A. S. and Muller, U. (2017). *Advancing Developmental Science*. London, UK: Routledge.
Dickson, M. (2006). "Plurality and complementarity in quantum dynamics." In Stephen H. Kellert, Helen E. Longino and C. Kenneth Walters (eds), *Scientific Pluralism: Minnesota Studies in the Philosophy of Science, Volume XIX*. Minneapolis, MN: University of Minnesota Press.
Dilthey, W. (1900). "The development of hermeneutics." In R. A. Makreel and F. Rodi (eds), *Wilhem Dilthey: Selected Works*. Volume 1. Princeton, NJ: Princeton University Press.
Dilthy, Dunbar, R. I. (1992). "Neocortex size as a constraint on group size in primates." *Journal of Human Evolution*. 20: 469–493.
Doctors, S. R. (2000). "Attachment-individuation: I. Clinical notes toward a reconsideration of 'adolescent turmoil'." *Adolescent Psychiatry*. 25: 3–16.
Dunbar, R. I. (1993). "Coevolution of neocortical size, group size and language in humans." *Behavioral and Brain Sciences*. 16: 681–735.
Durstewitz, D., Seamans, J. K. and Sejnowski, T. J. (2000). "Neurocomputational models of working memory." *Natural Neuroscience*. 3 (supplement): 1184–1191.
Dziegielewski, S. F. (2006). *Psycho-Pharmacology Handbook: For the Non-Medically Trained*. New York: W. W. Norton & Company.
Edelman, G. M. (1989). *The Remembered Present*. New York: Basic Books.
Einstein, A. (1991). In Schlipp, P. *Autobiographical Notes*. Peru, IL: Open Court Publishing.
Elkind, D. and Bowen, R. (January, 1979). "Imaginary audience behavior in children and adolescents." *Developmental Psychology*. 15: 38–44.
Engel, G. L. (April 8, 1977). "The need for a new medical model: A challenge for biomedicine." *Science*. New Series, 196(4286): 129–136.
Engel, G. L. (June, 1978). "The biopsychosocial model and the education of health professionals." *Annals of the New York Academy of Sciences*. Vol. 310. Primary Health Care in Industrialized Nations. 169–181.
Erdmann, J. E. (1866). *A History of Philosophy*. Fourth Edition. 1892. London, UK: Swan Sonnenschein & Co and New York: MacMillan & Co.
Erikson, E. H. (1950). *Childhood and Society*. Second edition. New York: W. W. Norton & Company.
Erikson, E. H. (1958). *Young Man Luther: A Study in Psychoanalysis and History*. New York: W. W. Norton and Company.
Fairbairn, W. R. D. (1941). "A revised psychopathology of the psychoses and psychoneuroses." *International Journal of Psychoanalysis*. 22: 250–279.
Fairbairn, W. R. D. (1952). *An Object Relations Theory of the Personality*. New York: Basic Books.
Federn, P. (1952). *Ego Psychology and the Psychoses*. New York: Basic Books.
Feinberg, T. E. (2009). *From Axons to Identity: Neurological Explorations of the Nature of the Self*. New York and London, UK: W. W. Norton & Company.
Ferenczi, S. (1988). *The Clinical Diary of Sandor Ferenczi*. Edited by J. DuPont. Cambridge, MA: Harvard University Press.

Fort Hood shooting, 2009. Wikipedia, no date. http://en.wikipedia.org/wiki/2009Fort_Hood_Shooting.

Fisher, H. E. (1998). "Lust, attraction, and attachment in mammalian reproduction." *Human Nature.* 9: 23–52.

Fraser, M. W., Kirby, L. D. and Smokowski, P. R. (2004). *Risk and Resilience in Childhood: An Ecological Perspective.* Washington, DC: National Association of Social Workers.

Fredrickson, B. L. (2001). "The role of positive emotions in positive psychology: The broaden-and-build theory of positive emotions." *American Psychologist.* 56. 93: 218–226.

Freud, S. (1891). *On Aphasia: A Critical Study.* Translated by E. Stengel. London, UK: Imago.

Freud, S. (1895). "Project for a scientific psychology." *Standard Edition. Vol. I,* pp. 283–397.

Freud, S. (1900). *The Interpretation of Dreams. Standard Edition. Vol. IV.*

Freud, S. (1905). "Three essays on the theory of sexuality." *Standard Edition. Vol. VII.* London, UK: Hogarth Press, pp. 125–245.

Freud, S. (1909). "Analysis of a phobia in a five year old boy." *Standard Edition. Vol. XX.* London, UK: Hogarth Press, pp. 3–149.

Freud, S. (1910). "The taboo of virginity." *Standard Edition. Vol. XI,* London, UK: Hogarth Press, p. 199.

Freud, S. (1915). "Papers on metapsychology." *Standard Edition. Vol. XIV,* London, UK: Hogarth Press, pp. 105–258.

Freud, S. (1920). "Beyond the pleasure principle: Revision of the theory of instincts." *Standard Edition. Vol. XVIII,* London, UK: Hogarth Press, pp. 34–64.

Freud, S. (1923). "The ego and the id." *Standard Edition. Vol. XIX,* London, UK: Hogarth Press, p. 15.

Freud, S. (1924). "Neurosis and psychosis." *Standard Edition. Vol. XIX,* London, UK: Hogarth Press, pp. 147–153.

Freud, S. (1933). "The question of a weltanschauung." Lecture XXXV. *Standard Edition. Vol. XXII,* London, UK: Hogarth Press, p. 158.

Freud, S. and Breuer, J. (1895). "Studies on hysteria." *Standard Edition. Vol. II.*

Friedman, L. (1999). *Identity's Architect: A Biography of Erik Erikson.* New York: Scribner.

Fu, L. (March 18, 2018). "Will millennials kill credit cards next?" Chicago, IL: *The Chicago Tribune.*

Gabbard, G. O. (1994). *Psychodynamic Psychiatry in Clinical Practice: The DSM-IV Edition.* Washington, DC The American Psychiatric Press, Inc., pp. 360–365.

Gabbard, G. O. (2005). *Psychodynamic Psychiatry in Clinical Practice.* Fourth Edition. Washington, DC: American Psychiatric Publishing.

Gabbard, G. O. (2014). *Psychodynamic Psychiatry in Clinical Practice.* Fifth Edition. Washington, DC: American Psychiatric Publishing.

Galatzer-Levy, R. M. and Cohler, B. J. (1993). *The Essential Other – A Developmental Psychology of the Self.* New York: Basic Books.

Galatzer-Levy, R. M. (2004). "Chaotic possibilities: Toward a new model of development." *The International Journal of Psychoanalysis.* 85: 419–441.

Galatzer-Levy, R. M. (2016). "The edge of chaos: A non-linear view of psychoanalytic technique." *International Journal of Psychoanalysis.* 97: 409–427.

Galatzer-Levy, R. M. (2017). *Nonlinear Psychoanalysis: Notes from Forty Years of Chaos and Complexity Theory.* London, UK and New York: Routledge.

Gay, P. (1988). *Freud: A Life for Our Time.* New York: W. W. Norton & Company.

Gedo, J. (1988). *The Mind in Disorder.* New York: Analytic Press.

Giammona, G. (October 5, 2015). "A mouse for millennial moms." Chicago, IL: *Chicago Tribune.*

References

Gilmore, K. (2008). "Psychoanalytic developmental theory: A contemporary reconsideration." *Journal of the American Psychoanalytic Association*. 58: 885–907.
Glinski, N. (March 2, 2015). "Millennials seem stuck amid gains." Chicago, IL: *Chicago Tribune*.
Glaser, G. (April, 2015). "The false gospel of Alcoholics Anonymous." *The Atlantic Monthly. God's Word: Today's Bible Translation That Says What It Means*. Grand Rapids, MI: World Publishing.
Goethe, J. W. (1774/1971). *The Sorrows of Young Werther*. New York: Vintage Books.
Goethe, J. W. (1821/1995). *Wilhelm Meister's Apprenticeship*. Eric A. Blackwell (ed.) Princeton, NJ: Princeton University Press.
Goldberg, A. (2001). *The Executive Brain: Frontal Lobes and Civilized Mind*. New York: Oxford University Press.
Goldberg, A. (2007). "Pity the poor pluralist." *Psychoanalytic Quarterly*. 765 (Supplement): 1663–1674.
Goldberg, A. (November 27, 2013). "A danger in diversity." Address delivered to the Chicago Psychoanalytic Society. Unpublished paper.
Goldstein, E. G. (1984). *Ego Psychology and Social Work Practice*. Second Edition. New York: The Free Press.
Greenberg, J. and Mitchell, S (1983). *Object Relations in Psychoanalytic Theory*. Cambridge, MA: Harvard University Press.
Greenman, L. (April, 2007). "Neuroscience and psychoanalysis: Approaches to consciousness and thinking." *Psychiatry*. 4(4): 51–57.
Greenspan, S. I. and Shanker, S. G. (2004). *The First Idea: How Symbols, Language and Intelligence Evolved from Our Primate Ancestors to Modern Humans*. Cambridge, MA: Da Capo Press/ Perseus Group.
Greiwe, E. (January 29, 2016). "Their dreams deferred: Millennials do want their own homes." Chicago, IL: *Chicago Tribune*.
Grondin, J. (2004). *Hans-Georg Gadamer: A Biography*. Translated by Joel Weinsheimer. New Haven, CT: Yale University Press.
Grosskurth, P. (1991). *The Secret Ring: Freud's Inner Circle and the Politics of Psychoanalysis*. Reading, MA: Addison-Wesley Publishing.
Guba, E. G. (1990). "The alternative paradigm dialogue." In E. G. Guba (ed.), *The Paradigm Dialogue*. Newbury Park, NY: Sage.
Habermas, J. (1969). *Knowledge and Human Interests*. Boston, MA: Boston Press.
Hadley, J. L. (1989). "The neurobiology of motivational systems." In J. Lichtenberg, *Psychoanalysis and Motivation*. Hillsdale, NJ: The Analytic Press, pp. 337–372.
Hadley, M. (2008). "Relational theory: Inside out, outside in and in-between and all around." *Inside Out and Outside In: Psychodynamic Clinical Theory and Psychopathology in Contemporary Multicultural Contexts*. Second Edition. New York: Jason Aronson.
Hall, G. S. (1904). *Adolescence: Its Psychology and its Relations to Physiology, Anthropology, Sociology, Sex, Crime, Religion, and Education* (two volumes). New York: Appleton.
Hanly, C. (1990). "The concept of truth in psychoanalysis." *International Journal of Psycho-Analysis*. 71: 375–383.
Harari, Y. N. (2015). *Sapiens: A Brief History of Humankind*. New York: HarperCollins.
Harris, A. (1991). "Gender as contradiction." *Psychoanalytic Dialogues*. 1: 197–220.
Harris, A. (2005). *Gender as Soft Assembly*. Hillsdale, NJ: The Analytic Press.
Harter, S. (2012). *The Construction of the Self: Developmental and Sociocultural Foundations*. Second Edition. New York and London, UK: The Guilford Press.
Hartmann, H. (1939). *Ego Psychology and the Problem of Adaptation*. New York: International Universities Press.

Harvard Mental Health Letter. (July, 2008). "Revisiting the CATIE schizophrenia study: Although questions remain, some clinical guidance has emerged." Boston, MA: The Harvard Medical School.
Harvard Mental Health Letter. (November, 2008). "Treating 'first-episode' schizophrenia: Current thinking about the best way to manage this critical phase." Boston, MA: The Harvard Medical School.
Harvard Mental Health Letter. (June, 2010). "Schizophrenia treatment recommendation Updated: The new PORT guidelines focus on improving physical as well as mental health." Boston, MA: The Harvard Medical School.
Haroutunian, S. (1983). *Equilibrium in the Balance: A Study of Psychological Explanation.* New York: Springer-Verlag New York, Inc.
Harvey, P. D. and Pinkham, A. (April, 2015). "Impaired self-assessment in schizophrenia: Why patients misjudge their cognitive functioning." *Current Psychiatry.* 14(4): 563–559.
Heidegger, M. (1962). *Being and Time.* Translated by John Macquarrie and Edward Robinson. New York: Harper & Row.
Heilman, R. M., Crisan, L. G., Houser, D., Miclea, M. and Miu, A. C. (2010). "Emotion regulation and decision making under risk and uncertainty." *Emotion.* 10(2): 257–265.
Hennessy-Fiske, M. (September 27, 2015). "For millennials, religious orders an unlikely calling." Chicago, IL: *Chicago Tribune.*
Herbling, D. (May 20, 2015). "Millennials want more on plate for breakfast." Chicago, IL: *Chicago Tribune.*
Herbling, D. (May 23, 2015). "Hostel checking into River North." Chicago, IL: *Chicago Tribune.*
Heuvel, K. (2015). "Can Hillary manage those rowdy populists?" *Washington Post.*
Hill, D. (2010). "Fundamentalist faith states: Regulation theory as a framework for the psychology of religious fundamentalism." Charles B. Strozier, David M. Terman, and James W. Jones with Katherine A. Boyd (eds), *The Fundamentalist Mindset.* New York: Oxford University Press.
Hinshelwood, R. D. (2013). *Research on the Couch: Single-Case Studies, Subjectivity and Psychoanalytic Knowledge.* New York: Routledge.
Hobson, J. A. and McCarley, R. W. (1977). "The brain as a dream-state generator." *American Journal of Psychiatry.* 134: 1335.
Hofer, M. A. (1995). "Hidden regulators: Implications for a new understanding of attachment, separation, and loss." In Goldberg, S., Muir, R. and Kerr, J. (eds), *Attachment Theory: Social, Developmental, and Clinical Perspectives.* Hillsdale, NJ: The Analytic Press, pp. 203–230.
Hofer, M. A. (2006). "Psychobiological roots of early attachment." *Association for Psychological Science.* 15(2): 84–88.
Hoffman, I. Z. (1983). "The patient as interpreter of the analyst's experience." *Contemporary Psychoanalysis.* 19: 389–422.
Hoffman, I. Z. (2002). *Ritual and Spontaneity in the Psychoanalytic Process.* Hillsdale, NJ: The Analytic Press.
Holy Bible. Revised Standard Version. Catholic Edition. (1966). New Testament. Charlotte, NC: Saint Benedict Press.
Hopfield, J. J. (1982). *Neural Networks and Physical Systems with Emergent Collective Computational Abilities.* Proc National Academy of Science USA, 79:2554- 2558.http:/en.wikipedia.org/wiki/Education_in_Ancient_Greece. 9/24/2014.
Hume, D. (1739/1740). *A Treatise of Human Nature.* A Public Domain Book.
Huppke, R. W. (November 16, 2015). "What do millennials, gen xers, and boomers share? Leadership values." Chicago, IL: *Chicago Tribune.*
Husserl, E. (1913/1950). *The Cartesian Meditations: An Introduction to Phenomenology.* Translated by Dorian Cairns. Springer Sciences + Business Media Dordrecht.

Hustvedt, S. (2016). *A Woman Looking at Men Looking at Women: Essays on Art, Sex, and the Mind*. New York: Simon and Schuster.

Imbasciati, A. (2017). *MindBrain, Psychoanalytic Institutions, and Psychoanalysts: A New Metapsychology Consistent with Neuroscience*. London, UK: Karnac.

Jackson, Don, D. (ed.). (1960). *The Etiology of Schizophrenia*. New York: Basic Books.

Jaffe, C. (2000). "Organizing adolescents(ce): A dynamic systems perspective on adolescence and adolescent psychotherapy." In C. Jaffe, *Adolescent Psychiatry*. Vol. 25. Hillsdale, NJ: The Analytic Press.

Jaffe, C. (2018). "Organizing adolescents(-ce) – A dynamic systems perspective on adolescence and adolescent psychotherapy." Issue VI. *Yellowbrick Journal of Emerging Adulthood*. Evanston, IL: Yellowbrick Foundation.

Janssen, K. (January 26, 2016). "Hilton targeting millennials with affordable hotel brand." Chicago, IL: *Chicago Tribune*.

Jay, M. (2017). *Supernormal: The Untold Story of Adversity and Resilience*. New York: Twelve, Hachette Book Group.

Johnson, O. A. (1992). "Immanuel Kant." In Ian P. McGreal (ed.), *Great Thinkers of the Western World*. New York: HarperCollins.

Jones, E. (20 February 2015). "Coming of age in Seoul." Agence-France Presse.

Kahn, A. Y., Zaidi, S. N., Salaria, S. N. and Bhattacharyya, A. (March, 2016). "Reducing morbidity and mortality from common medical conditions in schizophrenia: You play a vital role in systematic screening, initiating treatment, and maintaining follow-up." *Current Psychiatry*. 15(3): 30–40.

Kalman, B. (2008). *Mexico: The People*. New York: Crabtree Publishing Company.

Kandel, E. R. (2005). *Psychiatry, Psychoanalysis, and the New Biology of the Mind*. Washington, DC: American Psychiatric Publishing, Inc.

Kant, I. (1781/1787/2007). *Critique of Pure Reason*. Translated by Max Muller. Editor and Translator Marcus Weigelt. Suffolk, UK: Clays, Ltd.

Kaplan, L. J. (1984). *Adolescence: The Farewell to Childhood*. New York: Simon and Schuster.

Kaplan, Z. and Joseph, N. J. (2007). "Bar Mitzvah, Bat Mitzvah." In Michael Berenbaum and Fred Skolnik (eds), *Encyclopedia of Judaica*. Second Edition. Vol. 3. Detroit, MI: Macmillan Reference USA.

Kaplan-Solms, K. and Solms, M. (2002). *Clinical Studies in Neuro-Psychoanalysis: An Introduction to a Depth Neuropsychoanalysis*. Second edition. London, UK: Karnac Books.

Keilman, J. (November 11, 2015). "Parents, economy to blame for kids' failure to launch." Chicago, IL: *Chicago Tribune*.

Kellert, S. H., Longino, H. E. and Walters, C. K. (2006). *Scientific Pluralism: Minnesota Studies in the Philosophy of Science, Volume XIX*. Minneapolis, MN: University of Minnesota Press.

Kernberg, O. F. (1976b). "A psychoanalytic classification of character pathology." *Object Relations Theory and Clinical Psychoanalysis*. New York: Jason Aronson.

Kernberg, O. F. (2011). "Suicide prevention for psychoanalytic institutes and societies." *Journal of the American Psychoanalytic Association*. 60(4): 707–719.

Kirby, L. D. and Frazer, M. W. (1997). *Risk and Resilience in Childhood: An Ecological Perspective*. Washington, DC: National Association of Social Workers Press.

Klerman, G. L. (April, 1990). "The psychiatric patient's right to effective treatment: Implications of Osheroff v. Chestnut Lodge." *American Journal of Psychiatry*. 147: 4.

Knapton, S. (19 January 2018). "Adulthood now begins at 24." London, UK: *The Telegraph*.

Kohut, H. (1959). "Introspection, empathy, and psychoanalysis: An examination of the relationship between mode of observation and theory." *Journal of the American Psychoanalytic Association*. 7(3): 459–483.

Kohut, H. (1966/1978). "Forms and transformations of narcissism." *Journal of the American Psychoanalytic Association*, 14(2): 243–272.
Kohut, H. (1971). *The Analysis of the Self: A Systematic Approach to the Psychoanalytic Treatment of Narcissistic Personality Disorders*. New York: International Universities Press.
Kohut, H. (1977). *The Restoration of the Self*. New York: International Universities Press.
Kohut, H. (1984). *How Does Analysis Cure?* Chicago, IL: University of Chicago Press.
Kohut, H. (1991). "Four basic concepts in self psychology." In. P. H. Ornstein (ed.), *The Search for the Self: Selected Writings of Heinz Kohut. 1978–1981*, Vol. 4. Madison, CT: International Universities Press, pp. 447–470.
Kozulin, A. (1984). *Psychology in a Utopia: Toward a Social History of Soviet Psychology*. Cambridge, MA: Massachusetts Institute of Technology.
Kunkle, F. (September 6, 2015). "Pew: Millennials feel small in boomer wake." Chicago, IL: *Chicago Tribune*.
Lakoff, G. (2009). "The neural theory of metaphor." In R. Gibbs (ed.), *The Metaphor Handbook*. Cambridge, MA: Cambridge University Press.
Lakoff, G. and Johnson, M. (1980/2003). *Metaphors We Live By*. Chicago, IL: University of Chicago Press.
Lapham's Quarterly. New York: America Agora Foundation.
Leary, M. R. (2008). "Functions of the self in interpersonal relationships: What does the self actually do?" In J. V. Wood, A. Tesser and J. C. Holmes (eds), *The Self in Social Relationships*. New York: Psychology Press, pp. 95–137.
LeDoux, J. (2012). "Afterword." *Psychoanalytic Review*. 99: 594–606.
Leiser, B. M. (1992). "John Locke." In Ian P. McGreal (ed.), *Great Thinkers of the Western World*. New York: HarperCollins.
Levin, F. M. (2009). "Metaphor: A fascinating philosophical puzzle piece with neuropsychoanalytic (NP) implications." *Psychoanalytic Inquiry*, 29(1): 69–78.
Levin, F. M. (2015). "What neuropsychoanalysis teaches us about dreaming." *The Annual of Psychoanalysis*. Volume 38. Chicago Institute for Psychoanalysis. Astoria, NY: IPBooks, pp. 131–141.
Lewis, C. E. (January 4, 2016). "Can millennials regain faith in government?" www.socialjusticesolutions.org/2016/01/04/cam-millennials-regain-faith-government.
Lewis, R. D. (2003). "Cultural imperative: Global teams in the 21st century." Boston, MA: Intercultural Press, Inc.
Lichtenberg, J. D. and Hadley, J. L. (1989). *Psychoanalysis and Motivation*. Hillsdale, NJ: Analytic Press.
Lichtenberg, J., Lachmann, F. M. and Fosshage, J. L. (2011). *Psychoanalysis and Motivational Systems: A New Look*. New York: Routledge.
Liddle, P. F. (1987). "The symptoms of chronic schizophrenia: A re-examination of the positive-negative dichotomy." *British Journal of Psychiatry*. 151: 145–151.
Lieberman, M. D. (2007). "Social cognitive neuroscience: A review of core processes." *Annual Review of Psychology*. 58: 259–289.
Locke, J. (1689/1996). *An Essay Concerning Human Understanding*. Indianapolis, IN: Hackett Publishing Company, Inc.
Loh, M., Rolls, E. T. and Deco, G. (November 9, 2007). "A dynamical systems hypothesis of schizophrenia." *PLoS Computational Biology*, 3(11): e228.Doi:10.1371/journal.pcbi.0030228, Google Scholar.
Lucente, R. L. (2012). *Character Formation and Identity in Adolescence: Clinical and Developmental Issues*. Chicago, IL: Lyceum Books.

Luyten, P., Blatt, S. J. and Corveleyn, J. (2006). "Minding the gap between positivism and hermeneutics in psychoanalytic research." *Journal of the American Psychoanalytic Association.* 54: 571–610.

Lyons-Ruth, K. (1991). "Rapprochement or approchement: Mahler's theory reconsidered from the vantage point of recent research on early attachment relationships." *Psychoanalytic Psychology*, 8: 1–23.

MacLean, P. (1990). *The Triune Brain in Evolution: The Role of Paleocerebral Functions.* New York: Plenum Press.

Mahler, M. S., Pine, F. and Bergman, A. (1975). *The Psychological Birth of the Human Infant: Symbiosis and Individuation.* New York: Basic Books.

Makari, G. (2008). *Revolution in Mind: The Creation of Psychoanalysis.* New York: HarperCollins.

Mann, C. C. (February 6, 2015). "How humankind conquered the world." Book Review. New York: *The Wall Street Journal.*

Mann, L. (May 1, 2016). "The 'Great Delay': Many of today's 20 somethings are choosing to stay at home and live with their parents. How's that working out?" Chicago, IL: *Chicago Tribune.*

MarksJarvis, G. (August 2, 2015). "Dear millennials, you're curbing the economy. Move out." Chicago, IL: *Chicago Tribune.*

Mahler, M., Pine, F. and Bergman, A. (1975). *The Psychological Birth of the Human Infant.* New York: Basic Books.

Manchir, M. (August, 25, 2013). "Leaving nest gets expensive: Student –loan debt, high rents, and job market keeping young adults at home." Chicago, IL: *Chicago Tribune.*

MarksJarvis, G. (October 21, 2015). "Millennials, extended families could get mortgages." Chicago, IL: *Chicago Tribune.*

Marley, J. A. (1994/2004). *Family Involvement in Treating Schizophrenia: Models, Essential Skills and Processes.* London, UK and New York: Routledge.

Maroda, K. (1991). *The Power of Countertransference.* Northvale, NJ: Aronson.

Martin, J. (July 28, 2012). "Adults can like living with parents." Chicago, IL: *Chicago Tribune.*

Martin, E. J. (May 28, 2017). "Don't let boomerang kids stop you from downsizing." Chicago, IL: *Chicago Tribune.*

Martin, E. J. (October 15, 2017). "Live-in adult children can disrupt home sale." Chicago, IL: *Chicago Tribune.*

Maxwell, J. C. (2007). *The Maxwell Leadership Bible.* Nashville, TN: Maxwell Motivation, Inc.

Mayerowitz, S. (April 17, 2015). "For millennials, 'lifestyle hotels'." Chicago, IL: *Chicago Tribune.*

McCarley, R. W. and Hobson, J. A. (1975). "Neuronal excitability modulation over the sleep cycle: A structural and mathematical model." *Science.* 189: 58–60.

McCarley, R. W. and Hobson, J. A. (1977). "The neurobiological origins of psychoanalytic dream theory." *American Journal of Psychiatry.* 134: 1211–1221.

McGowan, K. (April, 2014). "The second coming of Sigmund Freud." *Discover Magazine*: Kalmbach Publishing Company.

McNamara, C. (September 14, 2011). "Full nests cause less of a flap: Stigma fades as more young adults stay home." Chicago, IL: *Chicago Tribune.*

McWilliams, N. (2005). "Preserving our humanity as psychotherapists." *Psychotherapy: Theory, Research, Practice, Training.* 42(2): 139–151.

McDaniel, K. (March 25, 2014). "Edith Stein: On the problem of empathy." Syracuse University, New York: Google Scholar.

McGowan, K. (April, 2014). "The second coming of Sigmund Freud." *Discovermagazine.com*

Mead, G. H. (1913/1964). "The social self." In Andrew J. Reck (ed.), *Selected Writings: George Herbert Mead*. Chicago, IL: University of Chicago Press.
Merleau-Ponty, M. (1945/2012). *Phenomenology of Perception*. Translated by Donald A. Landes. New York: Routledge.
Merleau-Ponty, M. (June 14, 2004). *Stanford Encyclopedia of Philosophy*. Stanford University.
Minuchin, S. (1974). *Families and Family Therapy*. Cambridge, MA: Harvard University Press.
Mitchell, S. A. (1988). *Relational Concepts in Psychoanalysis: An Integration*. Cambridge, MA: Harvard University Press.
Mitchell, S. A. (2000). *Relationality: From Attachment to Intersubjectivity*. Hillsdale, NJ: The Analytic Press.
Mitchell, S. A. and Black, M. J. (1995). *Freud and Beyond: A History of Modern Psychoanalytic Thought*. New York: Basic Books.
Mitchell, S. D. (2002). "Integrative pluralism." *Biology and Philosophy*. 17: 55–70.
Modell, A. (1997). "Reflections on metaphor and affects." *The Annual of Psychoanalysis*. 25: 219–233.
Modell, A. (2000). "The transformation of past experience." *The Annual of Psychoanalysis*. 28: 137–149.
Moll, H. and Tomasello, M. (2007). "Cooperation and human cognition: The Vygotskian intelligence hypothesis." *Philosophical Transactions of the Royal Society B*. London, UK. 362(1480): 639–648.
Monroe-Cook, D. (2016). "Ten years at Yellowbrick: What I learned." *Yellowbrick Journal*. Evanston, IL: Yellowbrick Foundation.
Montgomery, A. (2013). *Neurobiology Essentials for Clinicians: What Every Therapist Needs to Know*. New York and London, UK: W.W. Norton & Company.
Moore, J. (January, 11, 2015). "Gen Y turns to transit-friendly areas." Chicago, IL: *Chicago Tribune*.
Moss, M. E. (1992). *"Giambattista Vico." Great Thinkers of the Western World*. Edited by Ian P. McGreal. New York: Harper Collins Publishers.
Muller, R. A. (1992). "Gottfried Wilhelm Leibniz." In Ian P. McGreal (ed.), *Great Thinkers of the Western World*. New York: HarperCollins.
Mueser, K. T. and McGurk, S. R. (2004). "Schizophrenia." *Lancet*. 363: 2063–2072.
Muskal, M. (November 5, 2015). "Over half of black millennials know of police violence, report finds." Chicago, IL: *Chicago Tribune*.
Nasrallah, H. A. (December, 2013). "Repositioning psychotherapy as a neurobiological intervention." *Current Psychiatry*. 12(2): 18–19.
Nasrallah, H. A. (July, 2014). "Post-World War II psychiatry: 70 years of momentous change." *Current Psychiatry*. 13(7): 21–22, 49–50.
Nasrallah, H. A. (July, 2015). "Is There only one 1 neurobiologic psychiatric disorder with different clinical expressions? Can DSM-5 Disorders Be Consolidated Based on a Common Neurobiology?" *Current Psychiatry*. 14(7): 10–12.
Nelson, E. E. and Panksepp, J. (1998). "Brain substrates of infant-mother attachment: Contributions of opioids, oxytocin, and norepinephrine." *Neuroscience and Behavioral Reviews*. 22 : 437–452.
Newman, K. M. (1992). "Abstinence, neutrality gratification: new trends, new climates, new implications." *Annual of Psychoanalysis*. 20: 131–144.
Nietzsche, F. (1873/2010). "On truth and lie in a nonmoral sense." In *Nietzsche: On Truth and Untruth*. Translated and edited by Taylor Carman. New York: HarperCollins, pp. 15–49.
"Ninth Special Report to the US Congress on Alcohol and Health." (1997). *NIH Publication 97-4017*. Bethesda, MD: National Institute on Alcohol Abuse and Alcoholism, pp. 181–191.

Noy, P. (1977). "Metapsychology as a multimodal system." *International Review of Psychoanalysis*. 4: 1–12.
Offer, D. (1980). "Adolescent development: A normative perspective." In S. Greenspan and G. Pollack (eds), *The Course of Life. Vol. IV. Adolescence*. Madison, CT: International Universities Press.
O'Loughlin, M. (April, 2013). "Book review of Barratt's *What is Psychoanalysis?*" *Psychoanalytic Psychology*. 33(2).
Orange, D. M. (1995). *Emotional Understanding: Studies in Psychoanalytic Epistemology*. New York: Guilford Press.
Orange, D. M. (2013). "A Pre-Cartesian self." *International Journal of Psychoanalytic Self Psychology*, 8: 488–494.
O'Reilly, B. and Dugard, M. (2013). *Killing Jesus: A History*. New York: Henry Holt.
O'Reilly, R. C. (2006). "Biologically based computer models of high-level cognition." *Science* 314: 91–94.
Overton, W. F. (2015). "Processes, relations, and relational-developmental-systems." In W. F. Overton and P. C. M. Molenaar (Vol. eds) and R. M. Lerner (ed. in-chief), *Handbook of Child Psychology and Developmental Science. Vol. 1: Theory & Method* (7th. ed.). Hoboken, NJ: Wiley, pp. 9–62.
Oxford Classical Dictionary. (1996). Third Edition. Oxford and New York: Oxford University Press, pp. 528–529.
Packer, S. (June 28, 2012). "A belated obituary: Raphael J. Osheroff." *Psychiatric Times*, www.psychiatrictimes.com.
Page, C. (March 18, 2015). "Millennials not free of prejudice." Chicago, IL: *Chicago Tribune*.
Palombo, J. (1988). "Adolescent development: A view from self psychology." *Child and Adolescent Social Work Journal*. 5(3): 171–186.
Palombo, J. (1991). "Bridging the chasm between developmental theory and clinical theory: Part I. The Chasm and Part II. The Bridge." *The Annual of Psychoanalysis*. 19: 151–193.
Palombo, J. (1996). "The Diagnosis and treatment of children with learning disabilities." *Child and Adolescent Social Work Journal*. 13(4): 311–332.
Palombo, J. (2001). *Learning Disorders and Disorders of the Self. In Children and Adolescents*. New York: W.W. Norton.
Palombo, J. (2006). *Nonverbal Learning Disabilities: A Clinical Perspective*. New York: W. W. Norton.
Palombo, J. (2011). "Executive function conditions and self-deficits." Chapter 13. In Nina Rovineilli Heller and Alex Gitterman (eds), *Mental Health and Social Problems: A Social Work Perspective*. London, UK and New York: Routledge, Taylor & Francis Group.
Palombo, J. (2013a). "The self as a complex adaptive system, Part I: Complexity, metapsychology, and developmental theories." *Psychoanalytic Social Work*. 20(1): 1–15.
Palombo, J. (2013b). "The self as a complex adaptive system, Part II: Levels of analysis and the position of the observer." *Psychoanalytic Social Work*. 20(2): 115–133.
Palombo, J. (2016). "The self as a complex adaptive system, Part III: A revised view of development." *Psychoanalytic Social Work*. 23(2): 145–164.
Palombo, J. (2017a). "The self as a complex adaptive system, Part IV: Making sense of the sense of self." *Psychoanalytic Social Work*. 24(1): 37–53.
Palombo, J. (2017b). *The Neuropsychodynamic Treatment of Self-Deficits: Searching for Complementarity*. New York: Routledge.
Palombo, J., Bendicsen, H. and Koch, B. (2009). *Guide to Psychoanalytic Developmental Theories*. New York: Springer Press.

Panksepp, J. (1985). "Mood changes." In: P. Vinken, C. Bruyn, and H. Klawans (eds), *Handbook of Clinical Neurology*. Vol. 45. Amsterdam: Elsevier.

Panksepp, J. (1998). *Affective Neuroscience: The Foundations of Human and Animal Emotions*. New York: Oxford University Press.

Panksepp, J., Asma, S., Curran, G., Gabriel, R. and Grief, T. (12 August 2012). "The philosophical implications of affective science." *Journal of Consciousness Studies*. 19(3–4): 6–48.

Panksepp, J. and Biven, L. (2012). *The Archaeology of the Mind: Neuroevolutionary Origins of Human Emotions*. New York and London, UK: W.W. Norton and Co.

Pepper, S. C. (1942). *World Hypotheses: A Study in Evidence*. Berkeley, CA: University of California Press.

Perry, S., Cooper, A. and Michels, R. (May, 1987). "The psychodynamic formulation: Its purpose, structure, and clinical application." *American Journal of Psychiatry*. 144: 5, 543–550.

Phelan, M. (August, 20, 2017). "Millennials redefining collecting classic cars." Chicago, IL: *Chicago Tribune*.

Pine, F. (1990). *Drive, Ego, Object, & Self: A Synthesis for Clinical Work*. New York: Basic Books.

Plath, S. (1971). *The Bell Jar*. New York: Harper & Row Publishers, Inc.

Popper, K. (1935/2002). *The Logic of Scientific Discovery*. London, UK: Routledge Classics.

Porges, S. W. (1998). "Love: An emergent property of the mammalian autonomic nervous system." *Psychoneuroendocrinology*. 23: 837–861.

Porges, S. W. (2001). "The polyvagal theory: Phylogenetic substrates of a social nervous system." *International Journal of Physiology*. 42: 29–52.

Porges, S. W. (2003). "The polyvagal theory: Phylogenetic contributions to social behavior." *Physiology and Behavior*, 79: 503–513.

Porges, S. W. (2009). "Reciprocal influences between body and brain in the perception and expression of affect." In D. Fosha, D. J. Siegel and M. F. Solomon (eds) *The Healing Power of Emotion: Affective Neuroscience, Development and Clinical Practice*. New York: W.W. Norton & Company.

Prochnik, G. (August, 14, 2017). "The curious conundrum of Freud's persistent influence." A book review of F. Crews' *The Making of an Illusion*. New York: *New York Times*.

Psychiatry Online Guideline Watch. (2009).

Quigley, M. W. (June 17, 2016). "Getting adult kids on board for downsizing." *AARP* magazine.

Quillman, T. (2012). "Neuroscience and therapist self-disclosure: Deepening right brain to right brain communication between therapist and patient." *Clinical Social Work Journal*. 40: 1–9.

Quinceañera. Encyclopedia Britannica Online, no date. www.britannica.com/topic/quinceanera.

Rampell, C. (December 9, 2015). "Surprise: Millennials favor gun rights." Chicago, IL: *Chicago Tribune*.

Rangell, L. (1988). "The future of psychoanalysis: The scientific crossroads." *Psychoanalytic Quarterly*. 57: 313–340.

Rangell, L. (1997). "Into the second psychoanalytic century: One psychoanalysis or many. The unitary theory of Leo Rangell., MD." *Journal of Clinical Psychoanalysis*. 6: 451–612.

Rangell, L. (2002). "The theory of psychoanalysis: Vicissitudes of its evolution." *Journal of the American Psychoanalytic Association*. 50: 1109–1137.

Rapacon, S. (February 4, 2016). "6 Savings Tips for Millennials Who Want to Get Rich." *Newsweek*.

Raza, M., Hirapara, K. and Hussain, N. (April, 2016). "Be an activist to prevent edentulism among the mentally ill." *Current Psychiatry*. 15(4).

Reed, R. (July 20, 2017). "Hogs losing hold on young." Chicago, IL: *Chicago Tribune*.

Reis, B. (2011). "Reading Kohut through Husserl." *Psychoanalytic Inquiry*. 31: 75–85.
Ringstrom, P. A. (2010). "Meeting Mitchell's challenge: A comparison of relational psychoanalysis and intersubjectivity systems theory." *Psychoanalytic Dialogues*. 20: 196–218.
Rockmore, T. (2010). *Kant and Phenomenology*. Chicago, IL: University of Chicago Press.
Rolls, E. T. (2005). *Emotion Explained*. Oxford, UK: Oxford University Press, p. 606.
Rolls, E. T. and Deco, G. (2002). *Computational Neuroscience of Vision*. Oxford: Oxford University Press.
Rosenthal, P. (January 8, 2016). "Packaging punch: McDonald's unveils most visible piece of nearly yearlong face-lift." Chicago, IL: *Chicago Tribune*.
Rousseau, J. J. (1762/1979). *Emile or On Education*. Translated by A. Bloom. New York: Basic Books.
Rustin, J. (2013). *Infant Research and Neuroscience at Work in Psychotherapy: Expanding the Clinical Perspective*. New York and London, UK: W. W. Norton & Company.
Rutter, M. (2006). *Genes and Behavior*. Malden, UK: Blackwell.
Saleeby, D. (1994). "Culture, theory and narrative: The intersection of meaning in practice." *Social Work*. 39(4): 351–359.
Salinger, J. D. (1945). *The Catcher in the Rye*. Boston, MA: Little, Brown.
Sameroff, A. (1983). "Developmental systems: Context and evolution." In W. Kessen (ed.), *Mussen's Handbook of Child Psychology*. Vol. 1. New York: Wiley, pp. 237–394.
Samuels, A. (1989). "Analysis and pluralism: The politics of psyche." *Journal of Analytic Psychology*. 34: 33–51.
Sander, L. W. (2000). "Where are we going in the field of infant mental health?" *Infant Mental Health Journal*. 21: 5–20.
Sander, L. W. (2002). "Thinking differently: Principles of process in living systems and the specificity of being known." *Psychoanalytic Dialogues*, 12: 11–42.
Sasse, B. (2017). *The Vanishing American Adult*. New York: St. Martin's Press.
Schafer, R. (1979). "On becoming a psychoanalyst of one persuasion or another." *Contemporary Psychoanalysis*. 15: 345–360.
Schechter, K. (2014). *Illusions of a Future: Psychoanalysis and the Biopolitics of Desire*. Durham, NC and London, UK: Duke University Press.
Scheffler, I. (1982). *Science and Subjectivity*. Indianapolis, IN: Hackett.
Scheuerecker, J., Ufer, S., Zipse, M., Frodl, T., Koutsouleris, N., et. al. (2007). "Cerebral changes and cognitive dysfunctions in medication-free schizophrenia: An fMRI study." *Journal of Psychiatric Resources*. E-publication. PMID: 17559877.
Schimmel, P. (2014). *Sigmund Freud's Discovery of Psychoanalysis: Conquistador and Thinker*. London, UK and New York: Routledge.
Schore, A. N. (1994). *Affect Regulation and the Origin of the Self: The Neurobiology of Emotional Development*. Hillsdale, NJ: Erlbaum.
Schore, A. N. (1997a). "A century after Freud's project: Is a rapprochement between psychoanalysis and neurobiology at hand?" *Journal of the American Psychoanalytic Association*. 45(3): 807–840.
Schore, A. N. (1997b). "Interdisciplinary developmental research as a source of clinical models." In M. Moskowitz, C. Monk, C. Kaye and S. J. Ellman (eds), *The Neurobiological and Developmental Basis for Psychotherapeutic Intervention*. Northvale, NJ: Jason Aronson, pp. 1–72.
Schore, A. N. (2001a). "Effects of a secure attachment relationship on right brain development, affect regulation, and infant mental health." *Infant Mental Health Journal*. 22(1–2): 7–66.
Schore, A. N. (2001b). "Minds in the making: Attachment, the self-organizing brain, and developmentally-oriented psychoanalytic psychotherapy." *British Journal of Psychotherapy*. 17(3): 299–328.

Schore, A. N. (2002). "Advances in neuropsychoanalysis, attachment theory, and trauma research: Implications for self psychology." *Psychoanalytic Inquiry*. 22(3): 433–484.

Schore, A. N. (2003a). *Affect Dysregulation and Disorders of the Self*. New York: W. W. Norton & Co.

Schore, A. N. (2003b). *Affect Regulation and Repair of the Self*. New York: W.W. Norton & Co.

Schore, A. N. (2011). "The right brain implicit self lies at the core of psychoanalysis." *Psychoanalytic Dialogues*. 21:75–100.

Schore, A. N. (2017). "All our sons: The developmental neurobiology and neuroendocrinology of boys at risk." *Infant Mental Health Journal*. 38(1): 15–52.

Schrobsdorff, S. (December 14, 2015). "The millennial beard: Why boomers need their younger counterparts. And vice versa." New York: *Time Magazine*.

Seidman, L. J., Faracone, S.V., Goldstein, J. M., Goodman, J. M., Kremen, W. S., Toomey, R., et al. (1999). "Thalamic and amygdala-hippocampal volume reductions in first degree relatives of patients with schizophrenia: An MRI-based morphometric analysis." *Biological Psychiatry*. 46 : 941–954.

Shedler, J. (2010). "The efficacy of psychodynamic psychotherapy." *American Psychologist*. 65(2): 98–109.

Shedler, J. (2015). "Where is the evidence for 'evidence-based' therapy?" *The Journal of Psychological Therapies Primary Care*. 4: 47–59.

Shenton, M. E., Kikinis, R., Jolesz, F. A., Pollack, S. D., LeMay, M., Wible, C. G., et al. (1992). Abnormalities of the left temporal lobe and thought disorder in Schizophrenia: A quantitative magnetic resonance imaging study." *New England Journal of Medicine*. 327: 604–612.

Shergill, S. S., Bammer, M. J., Wilams, C. C., Murray, R. M. and McGuire, P. K. (2000). "Mapping auditory hallucinations in schizophrenia using functional magnetic resonance imaging." *Archives of General Psychiatry*. 57: 1033–1036.

Shields, D. and Salerno, S. (2013). *Salinger*. New York: Simon & Schuster.

Sell, S. S. (November 11, 2015). "Millennial women living at home at rate not seen since 1940." Chicago, IL: *Chicago Tribune*.

Sichelman, L. (January 31, 2016). "Why are so many young women living with parents?" Chicago, IL: *Chicago Tribune*.

Siegel, D. J. (1999). *The Developing Mind: How Relationships and the Brain Interact to Shape Who We Are*. New York and London, UK: Guilford Press.

Siegel, D. L. and Hartzel, M. (2003). *Parenting from the Inside Out: How a Deeper Self Understanding Can Help You Raise Children Who Thrive*. New York: Tarcher/Putnam.

Sirapada, B. and Jobe T. H. (October 25, 2017). "A biological topography for psychoanalysis." Presented at the Wednesday Psychoanalytic Symposium at the Chicago Institute for Psychoanalysis. Unpublished.

Sklansky, M. (1991). "The pubescent years: Eleven to fourteen." In Stanley I. Greenspan and George H. Pollock (eds), *The Course of Life. Vol. IV. Adolescence*. Madison, CT: International Universities Press.

Skurky, T. A. (1990). *The Levels of Analysis Paradigm: A Model for Individual and Systemic Therapy*. New York: Praeger.

Smaller, F. (2013). *The Psychoanalytic Vision: The Experiencing Subject, Transcendence, and the Therapeutic Process*. New York and London, UK: Routledge.

"Social Security chugs toward the cliff." (July 20, 2017). Editorial. Chicago, IL: *Chicago Tribune*.

Solms, M. (2000). "Dreaming and REM sleep are controlled by different brain mechanisms." *Behavioral and Brain Science*. 23: 843–850.

Solms, M. (2015). "*The interpretation of dreams* and the neurosciences." In M. Solms, *The Feeling Brain: Selected Papers on Neuropsychoanalysis*. London, UK: Karnac.
Solms, M. and Turnbull, O. H. (2011). "What is neuropsychoanalysis?" *Neuropsychoanalysis* 13(2): 133–146.
Spear, L. (2010). *The Behavioral Neuroscience of Adolescence*. New York: W. W. Norton.
Spencer, H. (1892). *The Principles of Psychology* (2 volumes). New York: D. Appleton & Company.
Spitz, R. A. (1945a). "Diacritic and coenesthetic organization: The psychiatric significance of a functional division of the nervous system into a sensory and emotive part." *The Psychoanalytic Review*. 32: 146–160.
Spitz, R. A. (1945b). "Hospitalism: An inquiry into the genesis of psychiatric conditions in early childhood." *Psychoanalytic Study of the Child*, 1: 53–74.
Spitz, R. A. (1965). *The First Year of Life: A Psychoanalytic Study of Normal and Deviant Development of Object Relations*. Madison, CT: International Universities Press, Inc.
Spitz, R. A. and Wolf, R. M. (1946). "The smiling response: A contribution to the ontogenesis of social relations." *Genetic Psychology Monograph*. 34: 57–125.
Sroufe, A. (1996). *Emotional Development: The Organization of Emotional Life in the Early Years*. New York: Cambridge University Press.
Stanford Encyclopedia of Philosophy. (June 14, 2004). "Maurice Merleau-Ponty." The *MetaphysicsResearch Lab*. Stanford, CA: Stanford University.
Stanford Encyclopedia of Philosophy. (December 16, 2013). "Phenomenology." The *MetaphysicsResearch Lab*. Stanford, CA: Stanford University.
Stanford Encyclopedia of Philosophy. (June 12, 2017). "Scientific realism." The *MetaphysicsResearch Lab*. Stanford, CA: Stanford University.
Stein, E. (1917/1989). *On the Problem of Empathy*. Translated by Waltrant Stein. Washington, DC: ICS Publications.
Stein, J. (May 20, 2013). "The ME ME ME Generation: Millennials are lazy, entitled narcissists who still live with their parents. Why they'll save us all." *Time*. 181(19).
Steinberg, L. (2005). "Cognitive and affective development in adolescence." *Trends in Cognitive Sciences*. 9(2): 69–74.
Steinberg, L. (2014). *Age of Opportunity: Lessons from the New Science of Adolescence*. Boston, MA and New York: Houghton Mifflin Harcourt.
Steinmetz, K. (October 26, 2015). "Help! My parents are millennials: How this generation is changing the way we raise kids." Tampa, FL: *Time Magazine*.
Stern, D. (1985). *The Interpersonal World of the Infant*. New York: Basic Books.
Stern, D. (1988). "The dialectic between the 'interpersonal' and the 'intrapsychic.'" *Psychoanalytic Inquiry* 8: 505–512.
Stern, D. B. (1997). *Unformulated Experience: From Dissociation to Imagination in Psychoanalysis*. Hillsdale, NJ: The Analytic Press.
Stern, D. B. (2013). "Psychotherapy is an emergent process: In favor of acknowledging hermeneutics and against the privileging of systematic empirical research." *Psychoanalytic Dialogues*. 23: 192–115.
Stevens, H. (May 28, 2017). "A good lecture from a GOP senator." Chicago, IL: *Chicago Tribune*.
Stiver, D. R. (1992). "Edmund Husserl." In Ian P. McGreal (ed.), *Great Thinkers of the Western World*. New York: HarperCollins.
Stolorow, R. D., Atwood, G. E. and Ross, J. (1978). "The representational world in psychoanalytic therapy." *International Review of Psychoanalysis*. 5: 247–256.
Stolorow, R. D. and Atwood, G. E. (1979/1993). *Faces in a Cloud: Intersubjectivity in Personality Theory*. New York: Jason Aronson.

Stolorow, R. D. and Atwood, G. E. (1983). "Psychoanalytic phenomenology: Progress toward a theory of personality." In A. Goldberg (ed.), *The Future of Psychoanalysis*. New York: International Universities Press, pp. 97–110.
Stolorow, R. D. and Atwood, G. E. (1992). *Contexts of Being: The Intersubjective Foundations of Psychological Life*. Hillsdale, NJ: The Analytic Press.
Stolorow, R. D., Orange, D. and Atwood, G. E. (2002). *Worlds of Experience: Interweaving Philosophical and Clinical Dimensions in Psychoanalysis*. New York: Basic Books.
Strenger, C. (1997). "Hedgehogs, foxes, and critical pluralism: The clinician's yearning for unified conceptions." *Psychoanalysis and Contemporary Thought*. 20: 111–145.
Subramanian, K., Kounios, J., Parrish, T. B. and Jung-Beeman, M. (2009). "A brain mechanism for facilitation of insight by positive affect." *Journal of Cognitive Neuroscience*. 21: 415–432.
Sulloway, F. J. (1979). *Freud: Biologist of the Mind – Beyond the Psychoanalytic Legend*. New York: Basic Books.
Sullivan, H. S. (1953). *The Interpersonal Theory of Psychiatry*. New York: W. W. Norton & Company.
Sullivan, H. S. (1962). *Schizophrenia as a Human Process*. New York: W. W. Norton & Company.
Sullivan, H. S. (1964). "Leadership, mobilization and postwar change." In H. S. Sullivan, *Fusion of Psychiatry and Social Science*. New York: W. W. Norton & Company.
Summers, F. (2006). "Fundamentalism, psychoanalysis, and psychoanalytic theories." *Psychoanalytic Review*. 93: 329–352.
Summers, F. (2013). *The Psychoanalytic Vision: The Experiencing Subject, Transcendence, and the Therapeutic Process*. New York and London, UK: Routledge.
Suth, A. B. (2011). Review of *Desire, self, mind and the psychotherapies: Unifying Psychological Science and Psychoanalysis*. (2008). In *Psychoanalytic Psychology*. Lanham, MD: Jason Aronson. 28: 154–161.
Taber's Cyclopedic Medical Dictionary. (1997). Philadelphia, PA: F. A. Davis Company.
Tangney, J. P. (2003). "Self-relevant emotions." In M. R. Leary and J. P. Tangney (eds), *Handbook of Self and Identity*. New York: Guilford Press.
Taylor, P. (February, 22–29, 2016). "The kids are all left: Millennials are financially stressed, socially liberal and politically pivotal. But none of that will matter if they don't vote." New York: *Time*.
Terman, D. (1988). "Optimum frustration: Structuralization and the therapeutic process." In A. Goldberg (ed.), *Progress in Self Psychology*. Vol. 4. Hillsdale, NJ: The Analytic Press, pp. 113–125.
Thelen, E. and Smith, L. B. (1994). *A Dynamic Systems Approach to the Development of Cognition and Action*. Cambridge, MA and London, UK: Massachusetts Institute of Technology.
The Way Way Back. (2013). Twentieth Century Fox Film Corporation and TSG Entertainment Finance LLC.
Thompson, G. (2000). *On Kant*. Belmont, CA: Wadsworth/Thompson Learning.
Tortora, G. J. (1986). *Principles of Human Anatomy*. New York: Harper and Row, Inc.
Tronick, E. (2007). *The Neurobehavioral and Social-Emotional Development of Infants and Children*. New York and London, UK: W. W. Norton & Company.
Tyson, P., Tyson, R. L. and Wallerstein, R. S. (1990). *Psychoanalytic Theory Development: An Integration*. Binghamton, NY: Vail-Ballou Press.
Tyson, P. (2002). "The challenges of psychoanalytic developmental theory." *Journal of the American Psychoanalytic Association*. 50: 19–52.
Tyson, P. (2004). "Points on a compass: Four views on the developmental theories of Margaret Mahler and John Bowlby." *Journal of the American Psychoanalytic Association*. 52: 499–509.

Urdang, E. (2008). *Human Behavior in the Social Environment: Interweaving the Inner and Outer Worlds*. New York and London, UK: Routledge.
Vaillant, G. E. (1983). *The Natural History of Alcoholism*. Cambridge, MA: Harvard University Press.
Waddell, M. (2018). *On Adolescence: Inside Stories*. London, UK and New York: Routledge.
Waelder, R. (1936). "The principle of multiple function." *Psychoanalytic Quarterly*. 5: 45–62.
Wagner, J. (November 1, 2015). "In Sanders, scarred millennials find suitable suitor." Chicago, IL: *Chicago Tribune*.
Walsh, F. (2003). "Family Resilience: Strengths forged through diversity." *Normal Family Processes: Growing Diversity and Complexity*. Third edition. New York: The Guilford Press.
Wang, X. J. (2001). "Synaptic reverberation underlying mnemonic persistent activity." *Trends in Neuroscience*. 24: 455–463.
Weber, L. (October 13, 2017). "Can the millennials put an end to sexual harassment?" Chicago, IL: *Chicago Tribune*.
Webster's New Collegiate Dictionary. (1981). Springfield, MA: G. and C. Merriam Co.
Weakland, J. H. (1960). "The 'double bind' hypothesis of schizophrenia and three party interaction." In *The Etiology of Schizophrenia*. New York: Basic Books, pp. 373–388.
Wedding, D. and Corsini, R. J. (2005). *Case Studies in Psychotherapy*. Fourth edition. Belmont, CA: Brooks/Cole – Thomson Learning.
Werner, E. E. and Smith R. S. (2001). *Journeys from Childhood to Midlife: Risk, Resilience and Recovery*. Ithaca, NY: Cornell University Press.
Whitaker, R. (2010). *Anatomy of an Epidemic: Magic Bullets, Psychiatric Drugs and the Astonishing Rise of Mental Illness in America*. New York: Broadway Paperbacks.
Wilhelm, H. (May 26, 2017). "Bringing adulthood back." Chicago, IL: *Chicago Tribune*.
Williams, C. (October 25, 2015). "From ugly to cool: Automotive outcasts find favor with millennials," Chicago, IL: *Chicago Tribune*.
Winnicott, D. W. (1964). *The Child, the Family and the Outside World*. Harmondsworth, UK: Penguin.
Winnicott, D. W. (1971). "The use of the object and relating through identifications." *Playing and Reality*. London, UK: Tavistock.
Winterer, G., Coppola, R., Goldberg, T. E., Egan, M. F., Jones, D. W., et al. (2004). "Prefrontal broadband noise, working memory, and genetic risk for schizophrenia." *American Journal of Psychiatry*. 161: 490–500.
Winterer, G., Musso, F., Beckman, C., Mattay, V., Egan, M. F., et al. (2006). "Instability of prefrontal signal processing in schizophrenia." *American Journal of Psychiatry*. 163: 1960–1968.
Winterer, G., Ziller, M., Dorn, H., Frick, K., Mulert, C., et al. (2000). "Reduced signal-to-noise ratio and impaired phase-locking during information processing." *Clinical Neurophysiology*. 111: 837–849.
Witherington, D. C. (2018). "Dynamic systems theory." In Anthony Steven Dick and Ulrich Muller (eds), *Advancing Developmental Science: Philosophy, Theory and Method*. London, UK: Routledge, Chapter 4. pp. 41–52.
Wolf, E. (1988). *Treating the Self: Elements of Clinical Self Psychology*. New York and London, UK: The Guilford Press.
Wong, G. (October 28, 2015). "Attic as poor man's penthouse: Fierce rental market leads landlords to convert unused space for tenants." Chicago, IL: *Chicago Tribune*.

INDEX

Note: Page numbers in **bold** denote tables. End of chapter notes are denoted by a letter n between page number and note number.

Abraham, K. 8
addiction 40n1; *see also* alcoholism
adolescent coming-of-age rituals 142–143
adolescent phase, prolongation of 143–147, 155
adversarial selfobjects 113–114
Agosta, L. 75
Alcoholics Anonymous (AA) 50, 53, 56, 58n1
alcoholism 50–51, 53, 56, 58n1; theories of 160–163
Aleman, A. 100
Alexander, F. 9
algorithm: definition of 66, 85; *see also* developmental algorithm
alter-ego selfobjects 113–114
ambivalence 114–115
Ammaniti, M. 153
amygdala 26, 29, 30, 31, 99
anterior cingulate cortex 30, 40n1, 100
antipsychotic medications *see* psychotropic medications
apperception 72
Arnett, J. 143–144, 155
Aron, I. 76
Aserinsky, E. 18n4
attachment theory 10, 24, 66, 90–92, 115
attachment-individuation process 90–91, 112
attractors 98, 111, 118, 120, 124n1

Atwood, G. E. 36, 69, 113, 114, 115, 116, 151
autonomic nervous system 29–30, 38, 132–133

Bacal, H. 97
Barratt, B. B. 4
Belsha, K. 85n1
Benjamin, J. 76, 78, 116–117
biomedical model 32
biopsychosocial model 32–33
Black, M. J. 22, 69
Blitzsten, L. 9
Bloom, A. 61–62
Blos, P. 36, 85, 89, 148
Bollas, C. 48–49, 76
bootstrapping 84
Bosanquet, B. 59–60
Bowen, A. 146
Bowen, R. 114
Bowlby, J. 105
brain development and organization 24–27, 150
Brazelton, T. B. 33
Brentano, F. 70
Breuer, J. 7, 8
Broca, P. P. 6
Bromberg, P. M. 76, 96
Brooks, D. 61
Bucci, W. 20n5

case formulation 88–124, 128–137; attachment theory 90–92; cognitive theory 98–100; complexity theory 98, 118–120; modern metaphor theory 88–90; neurobiology with narrative theory 121–124; non-linear dynamic systems perspective 134–136; self psychology with intersubjectivity theory and relational psychoanalysis 93–98; single theory 83; social brain perspective 132–134, 136; supportive relationships 96–98; theoretical considerations 80–85; traditional psychodynamic perspective 129–132; *see also* contemporary psychoanalytic developmental theory
case study (Myles) 42–58, 151–152, 156; alcoholism 50–51, 53, 56, 58n1; attenuation of grandiosity 52, 57, 112–114; cognitive dissonance 99; cognitive theory 98, 99, 100; complexity theory 98, 120; decision making under emotional strain 100; dream session 54–55; neurobiology with narrative theory 121, 123–124; non-linear dynamic systems perspective 136; psychotropic medications 45, 46, 50, 54, 55, 56, 141n1; responses to therapeutic situation 131–132; secure attachment 92, 123–124, 130; selfobjects 47, 48, 49, 52–53, 54, 56, 58n1, 92, 93, 98, 131–132; self-referencing metaphors 44, 46, 51, 52, 53–54, 57, 58, 88, 89, 96, 99, 112–113, 120, 123, 131, 151, 156; sense of self 93–96; social brain perspective 132–134; traditional self psychology formulation 129–132; treatment plan 50, 58, 120, 133, 139
cerebral cortex 25, 26; cingulate cortex 26, 30, 40n1, 100; insula cortex 26, 40n1; prefrontal cortex 2, 26, 28, 29, 30, 31, 62, 99, 123, 132–133, 135, 150
chaos, edge of 119, 120
Charcot, J. M. 6
Chew, G. 84
Chicago Institute for Psychoanalysis (CIP) 9
cingulate cortex 26, 30, 40n1, 100
Ciompi, L. 27, 28–29
Claridge, G. 94–95
Cloninger, C. R. 155
co-created meaning-making 33, 35, 36, 115
cognition 24, 66, 98–100; hot and cold 99
cognitive conceit 114
cognitive dissonance 99, 112, 114–115
cognitive reappraisal 100
Cohler, B. J. 97–98

Colarusso, C. A. 85, 115, 148
cold cognitions 99
coming-of-age rituals 142–143
complex systems 37–38
complexity theory 24, 33, 47–48, 63, 66, 98, 118–120
contemporary psychoanalytic developmental theory 24, 66, 101–118; developmental models 101–111, **110**; developmental processes 111–118, 153–154
contextualism 68, 87n4
Cooper, A. 4–5, 10–11
Corsini, R. J. 128
coupled oscillators 120
Cozolino, L. 25, 26, 28, 29–31, 64–65, 66, 94, 95, 105–106, 111, 121–123, 133, 150
critical pluralism 87n5
critical thinking mental health decision-making flow chart 42, 159
Curtis, R. C. 11–12, 20n5

Dahl, R. E. 31
Damasio, A. R. 62–63, 121, 151
Davies, J. M. 76
decision making: critical thinking mental health decision-making flow chart 42, 159; under emotional strain 99–100
Deresiewicz, W. 61–62
Descartes, R. 71, 80, 81
developmental algorithm 14, 66–68, 82, 84, 85, 88–124, 149, 152–153; attachment theory 66, 90–92; cognitive theory 66, 98–100; complexity theory 66, 98, 118–120; definition of algorithm 66, 85; elements of 66–67, 152; modern metaphor theory 66, 88–90; neurobiology with narrative theory 67, 121–124; self psychology with intersubjectivity theory and relational psychoanalysis 66, 93–98; supportive relationships 96–98; *see also* contemporary psychoanalytic developmental theory
developmental models 101–111, **110**
developmental pathways 59–62, 152
developmental processes 111–118, 153–154
developmental science 67–68
Diagnostic and Statistical Manual of Mental Disorders 9, 100
Dick, A. S. 153
Dickson, M. 104
Dilthey, W. 76
Doctors, S. R. 90–91

dorsal anterior cingulate cortex 40n1
dreaming 18n4, 54–55
drive theory 6–8, 77, 102, 110
Dugard, M. 126n5
Durstewitz, D. 135
dynamic localization 13
dynamic systems theory 33, 35, 47–48, 68, 115; *see also* non-linear dynamic systems theory

eclecticism 22, 34, 83, 84, 102–103
edge of chaos 119, 120
education, ancient Athens 59–60
ego allies 97, 98
ego functions 96–97
ego psychology 9, 24, 36, 96–97, 148
egocentricity 114
ego/self ideal 118
Einstein, A. 156
Elkind, D. 114
embodiment of metaphor 88–90
emergence 68, 118, 119–120
emerging adulthood concept 143–144, 155
emotional strain, decision making under 99–100
empathy 74–75, 78, 93
empiricism 70–72, 82
Engel, G. L. 32–33
ephebic training 59–60
epigenetic developmental framework 28–32, 150–151
epigenetic hierarchical arrangement 84–85
Erikson, E. H. 144, 155, 156
essential others 97–98
executive functioning 2, 28, 63–64, 79, 85n1, 122, 150
expressive suppression 100

family psychosocial services 139, 140
fear regulation system 28, 29
Federn, P. 137n1
Feinberg, T. E. 63, 121, 122, 151
Ferenczi, S. 8
Fisher, H. E. 30
Fliess, W. 7, 8
formulation *see* case formulation
Fraser, M. W. 37
Fredrickson, B. L. 39
Freud, A. 8–9
Freud, S. 6–8, 12–14, 15, 83, 90, 101, 102, 110, 137n1, 150, 156
futurity 116, 124n2

Gabbard, G. O. 155
Gadamer, H.-G. 70, 76

Galatzer-Levy, R. M. 47, 97–98, 118, 119–120
Gallese, V 153
Ganzer, C. 78
Gedo, J. 84–85
gender differences 38, 39–40
Generation Y *see* millennials
Gilmore, K. 101–102
Girardi, L. 85n1
Goethe, J. W. 60, 142
Goldberg, A. 3–4, 79, 136–137
Goldstein, E. G. 97
grandiosity, attenuation of 52, 57, 112–114
Greece, ancient 59–60
Greenspan, S. I. 107, 111

Habermas, J. 35
Hadley, J. L. 20n6, 108, 110–111
Hadley, M. 70, 77
Hall, G. S. 147
hallucinations 46, 49, 95, 130, 135, 136
Hanly, C. 5
Harari, Y. N. 124n2
hard sciences 5
Haroutunian, S. 86n2
Harris, A. 76
Harter, S. 28, 31–32, 114–115, 150–151
Hartmann, H. 8
Harvard Mental Health Letter (HMHL) 138, 139
Heidegger, M. 69, 70, 75, 80
Heilman, R. M. 100
hermeneutics 76, 82
hidden regulators 27, 28, 150
Hill, D. 65–66
Hinshelwood, R. D. 5
hippocampus 27
Hobson, J. A. 18n4
Hofer, M. A. 27, 28, 150
Hoffman, I. Z. 76, 97
hot cognitions 99
HPA (hypothalamic-pituitary-adrenal axis) complex of hormone regulation 28, 29, 150
Hume, D. 71–72, 80
Husserl, E. 69, 70, 72–74, 75, 80
Hustvedt, S. 80–81, 89, 90
hypothalamus 27

ideal self 118
idealization 93, 131–132
Imbasciati, A. 15–17
individuation 90–92, 112; *see also* separation-individuation theory
infant-caregiver interaction 27, 28, 33, 34–36, 105, 106, 107, 115, 151

insecure attachment 37, 65
insight problem solving 100
insula cortex 26, 40n1
integration of theoretical orientations, definition of 82
internal dialogue 122–123
intersubjectivity theory 24, 33, 35, 66, 68–76, 77–79, 93–98, 115, 116–117, 153

Jaffe, C. M. 23–24, 84–85, 119
James, W. 61
Jay, M. 39
Jesus 126n5
Jobe T. H. 17
Johnson, M. 89
Johnson, O. A. 72, 77
Jung, C. 6

Kahn, A. Y. 139
Kandel, E. R. 12–13
Kant, I. 70–71, 72, 75, 80
Kaplan-Solms, K. 6, 7, 8, 13, 14
Kernberg, O. F. 9
Kirby, L.D. 37
Klein, M. 8–9
Kleitman, N. 18n4
Knapton, S. 147
Kohut, H. 9, 69, 91, 93, 97, 110, 111
Kozulin, A. 13
Kunkle, F. 147

Lakoff, G. 89, 110
language processing 122–123
Lapham's wound metaphor 61–62, 152
Leary, M. R. 32
LeDoux, J. 155
Leibniz, G. W. 71
Levin, F. M. 18n4
Lichtenberg, J. D. 20n6, 89, 107–108, 110–111
Lieberman, M. D. 99
limbic system 25, 26
Locke, J. 71
Loh, M. 134–135
Lucente, R. L. 153
Luke, Gospel of 126n5
Luria, A. R. 13
Luyten, P. 11
Lyons-Ruth, K. 90–91

McCarley, R. W. 18n4
McDaniel, K. 74–75
MacLean, P. 24–25
Mahler, M. S. 9, 90–91, 111, 116, 148

Marley, J. A. 140
Maroda, K. 76
Mead, G. H. 76
medical model of brain disorders 9–10
medications see psychotropic medications
Merleau-Ponty, M. 75–76, 80, 90, 110
metaphor theory, modern 24, 66, 88–90
metaphors: embodiment of 88–90; Lapham's wound 61–62, 152; root 87n4; self-regulating metaphor 2, 112, 123, 152, 154; see also self-referencing metaphor
metapsychology 7, 14, 15–17
millennials 61, 144–147, 155
mindsharing 106–107
mirroring 93, 131
Mitchell, S. A. 22, 69, 76, 77
Mitchell, S. D. 104
Modell, A. 89
modern metaphor theory 24, 66, 88–90
Moll, H. 25
monism 13–14, 103
motivation: complexity theory and 98; Cozolino's social brain theory of 28, 29–31, 111; embodiment of metaphor and 89; Freud's conceptualization of 102; Lichtenberg's theory of 20n6, 107–108; Panksepp's theory of 89, 108–111, **110**
MRM see Mutual Regulation Model (MRM)
Muller, U. 153
mutual recognition, differentiation of 32, 116–117, 123
Mutual Regulation Model (MRM) 33–36, 38, 115, 151
Myles see case study (Myles)

narcissism 93, 110
narrative theory 24, 67, 121–124
Nasrallah, H. R. 14–15, 40n1
Nelson, E. E. 30
neural self 62–63, 112, 121, 122, 151
neurobiological regulation 1, 27–31
neurobiology with narrative theory 24, 67, 121–124
neurochemicals 30
neuroendocrinology regulatory hypothesis of boys at risk 39–40
neurology 6–7
neurons 6
neuropsychoanalysis 12–15, 150
neuropsychoanalytic metapsychology 15–17
neuroscience 10; phenomenology and 74; psychoanalysis and 10–11, 12–15, 18n4, 150

Newman, K. M. 97
Nietzsche, F. 88
non-linear dynamic systems theory 10, 24, 33, 47–48, 77, 98, 118–120; case formulation 134–136
normal stress resilience hypothesis 37, 38–39
normative dissonance 114–115
Noy, P. 22

object relations theory 9, 69, 116
O'Loughlin, M. 4
orbital medial prefrontal cortex 26, 29, 30
O'Reilly, B. 126n5
organicism 68, 87n4
Osheroff, R. J. 9, 18n3
Overton, W. F. 67–68

Palombo, J. 5, 6, 12, 23, 33, 67, 68, 77, 79, 81–82, 87n4, 91–92, 93, 95, 103, 104–107, 114, 117, 119, 126n6, 130–131, 150
Panksepp, J. 18n4, 30, 89, 107, 108–111, **110**, 122
parasympathetic autonomic nervous system 30, 38, 132
Pepper, S. C. 68, 87n4
Perry, S. 129–132
perspectival realism 76–77
phenomenology 69, 70–76
Piaget, J. 98
Pine, F. 22
Plato 71
pluralism 12, 83–84, 87n5, 103–104, 136–137
polyvagal theory of social engagement 30, 133
Popper, K. 156–157
Porges, S. W. 30, 133
positivism 80–82, 103, 104, 125n3
pragmatism 87n5
prefrontal cortex 2, 26, 28, 29, 30, 31, 62, 99, 123, 132–133, 135, 150
Project for a Scientific Psychology (Freud) 7, 8, 12, 13, 15, 150
protective factors 37
psychoanalytic theories 3–12, 149–150; *see also* contemporary psychoanalytic developmental theory; neuropsychoanalysis
psychobiological co-regulation 33–36, 151
Psychodynamic Formulation Model 129–132
psychoeducational services 139, 140
psychosexual synthesis 8, 101

psychosocial moratorium concept 144
psychotropic medications 138, 141n1; case study 45, 46, 50, 54, 55, 56, 141n1
puberty 29, 31
purism 87n5

Quillman, T. 30

Rangell, L. 83–84
Rapacon, S. 146
rationalism 70–72
realism: perspectival 76–77; scientific 12, 82, 103
reality, sense of 113, 114–115
reflexive social language 122, 123
regulation theory 1, 22–40, 148, 149, 150–151; biopsychosocial model 32–33; brain development and organization 24–27, 150; Cozolino's model 28, 29–31, 150; definition of 65–66, 86n2; epigenetic developmental framework 28–32, 150–151; fear regulation system 28, 29; gender differences 38, 39–40; Harter's model 28, 31–32, 150–151; Hofer's model 27, 28, 150; Mutual Regulation Model (MRM) 33–36, 38, 115, 151; neurobiological regulation 1, 27–31; neuroendocrinology regulatory hypothesis of boys at risk 39–40; resilience 37–39, 151; Schore's model 27, 28, 150; self-state regulation 31–32; Siegel's model 27–29, 150; social engagement system 28, 29–30, 133; social motivation system 28, 30; stress regulation system 28, 29; *see also* developmental algorithm
Reis, B. 69
relational psychoanalysis/theory 24, 66, 67, 68, 76–79, 93–98, 153
REM sleep 18n4
resilience 37–39, 151
reward representation and reinforcement 30
right brain-left brain relationship 94–95
Ringstrom, P. A. 76–77, 78
risk factors 37
risk taking 31, 150
Rolls, E. T. 135
root metaphors 87n4
Rousseau, J. J. 147
Rustin, J. 10
Rutter, M. 37

Samuels, A. 84
Sander, L. W. 106
Sartre, J.-P. 70

Sasse, B. 147
Schafer, R. 23
Schechter, K. 9, 97
Scheuerecker, J. 135
Schimmel, P. 90
schizophrenia 40n1, 100, 137n1; comorbid conditions 139; dynamical systems hypothesis of 134–136; psychosocial interventions 139–140; psychotropic medications 138, 141n1; treatment of 138–140, 154–155; *see also* case study (Myles)
Schleiermacher, F. 76
Schore, A. N. 27, 28, 39–40, 63, 95, 105, 121, 150, 151
Schrobsdorff, S. 146
scientific pluralism 12, 103–104
scientific realism 12, 82, 103
Secret Committee 8
secure attachment 37, 65, 92, 123–124, 130
self: as a complex adaptive system 91–92, 103, 104–107, 121, 151; sense of 92, 93–96; theory of origins of 109–110
self psychology 9, 24, 66, 69, 79, 93–98, 151, 153; traditional case formulation 129–132
self-conscious emotions 32, 150–151
self-delineating selfobjects 36, 113–114, 115, 151
selfobjects 39, 67, 69, 91, 97, 98, 112; adversarial 113–114; alter-ego 113–114; attenuation of grandiosity and 113–114; case study 47, 48, 49, 52–53, 54, 56, 58n1, 92, 93, 98, 131–132; definition of 78–79; developmental needs 93, 131–132; self-delineating 36, 113–114, 115, 151
self-referencing metaphor 2, 89–90, 112, 114, 118, 123, 149, 151, 152, 154; case study 44, 46, 51, 52, 53–54, 57, 58, 88, 89, 96, 99, 112–113, 120, 123, 131, 151, 156; in Gospel of Luke 126n5
self-reflection 122, 123
self-regulating metaphor 2, 112, 123, 152, 154
self-similarity 120
self-state regulation 31–32
sense of reality 113, 114–115
sense of self 92, 93–96
separation-individuation theory 10, 90–91, 111, 116
Shanker, S. G. 107, 111
Shedler, J. 11–12
Shergill, S. S. 135
Siegel, D. J. 27–29, 64, 66, 118, 122, 124n1, 150

single theory formulations 83
Sirapada, B. 17
Skurky, T. A. 119
Smith, L. B. 111, 124n1
Smith R. S. 37
social brain hypothesis 25–27, 150; case formulation 132–134, 136; Cozolino's theory of motivation 28, 29–31, 111
social constructivism 67, 76, 77, 80–82, 97–98, 125n3
social engagement system 28, 29–30, 133
social motivation system 28, 30
soft sciences 5–6
Solms, M. 6, 7, 8, 13, 14, 17, 18n4
somatosensory cortex 26
Sorrows of Young Werther, The (Goethe) 60
Spear, L. 63–64
Spencer, H. 86n2
Spitz, R. A. 9
Sroufe, A. 27–28
state of mind 122, 123
Stein, E. 70, 74–75
Steinberg, L. 63, 85n1
Steinmetz, K. 146
Stern, D. B. 9, 76, 105, 116
Stevens, H. 147
still face experiment 38
Stolorow, R. D. 36, 69, 76–77, 113, 114, 115, 116, 151
strange attractors 98, 124n1
Strenger, C. 87n5
stress regulation system 28, 29
Subramanian, K. 100
Sullivan, H. S. 76, 115, 137n1
Sulloway, F. J. 12
Summers, F. 24, 79, 116, 124n2
supportive relationships 96–98; *see also* selfobjects
sympathetic autonomic nervous system 29, 30, 38, 132, 133

Tangney, J. P. 32
Terman, D. 97
Thelen, E. 111, 124n1
theory building 5–6
Thompson, G. 70–71
Tomasello, M. 25
"Top dog on the floor" self-referencing metaphor 46, 51, 52, 53–54, 88, 89, 96, 99, 112–113, 120, 123, 131, 151, 156
transcendental unity of apperception 72
transference 93, 131–132
Transformational Self concept 2, 62–64, 112, 121, 125n4, 148–149, 151–152, 156
triune brain hypothesis 24–25, 150

Tronick, E. 33–36, 37–39, 102–103, 115, 151
Turnbull, O. H. 13
twinning 93, 132
Tyson, P. 4, 9, 10, 102

Urdang, E. 37
usable objects 97, 98

vagal brake 30, 133
vagal system of autonomic regulation 29–30, 133
ventral striatum 30
Vico, G. 80–81
Viner, J. 92

Waddell, M. 60–61
Walsh, F. 37
Wang, X. J. 135
Weakland, J. H. 137n1
Wedding, D. 128
Werner, E. E. 37
Wernicke, C. 6
Whitaker, R. 141n1
Winnicott, D. W. 35, 97, 116
Witherington, D. C. 67–68
Wolf, E. 54, 78–79, 91, 93, 113
Wong, G. 145
working memory
    dysfunction 135

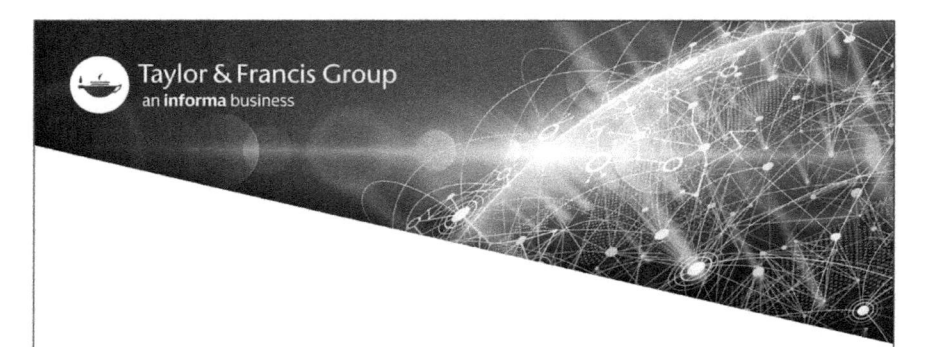

Taylor & Francis eBooks

www.taylorfrancis.com

A single destination for eBooks from Taylor & Francis with increased functionality and an improved user experience to meet the needs of our customers.

90,000+ eBooks of award-winning academic content in Humanities, Social Science, Science, Technology, Engineering, and Medical written by a global network of editors and authors.

TAYLOR & FRANCIS EBOOKS OFFERS:

- A streamlined experience for our library customers
- A single point of discovery for all of our eBook content
- Improved search and discovery of content at both book and chapter level

REQUEST A FREE TRIAL
support@taylorfrancis.com

Routledge
Taylor & Francis Group

CRC Press
Taylor & Francis Group